Book of Nations

**The World Book
Desk Reference Set**

Book of Nations

Published by

World Book Encyclopedia, Inc.
a Scott Fetzer company

Chicago

Staff

Contents

Introduction

There is a world full of nations—warm and chilly, large and little, ancient and even brand-new. In all, there are 168 independent countries you should know about. *The World Book of Nations* presents them all.

This convenient little handbook enables you to locate all sorts of information about the world's nations by quickly flipping through the pages. First come the entries for individual countries. These are numbered and arranged alphabetically by name. The entries give a country's geographic location, plus statistics such as area, population size, largest cities, climates, and official name (if different from the name by which the country is listed). The climates for each nation are named. Last comes a profile of the country, explaining what kind of people live there, where they came from, and how they make their living.

The section titled "Statistics and maps," which follows the 168 nation entries, shows the nations by continent. The maps will help you pinpoint any nation on the globe. The numbers on the maps correspond to the numbers assigned to the individual nation entries that comprise the first part of the volume.

Afghanistan

Afghanistan (1) is located in south-western Asia. It is bordered by the Soviet Union, China, Pakistan, and Iran.

Capital: Kabul
Official Name: In Pashto, Da Afghanistan Democratic Jamhouriat; In Dari, Jamhouriat-i-Democratic-i-Afghanistan (Democratic Republic of Afghanistan)
Official Languages: Pashto (also called Pushtu) and Dari
Government: Republic—cabinet responsible to Revolutionary Council of military officers and civilians
National Anthem: "Sorode Jamhouriat" ("Anthem of the Republic")
Flag: One horizontal stripe each of black, red, and green superimposed with Afghanistan's coat of arms
Area: 250,000 sq. mi. (647,497 km^2); greatest distances—east-west, 820 mi. (1,320 km); north-south, 630 mi. (1,012 km)
Population: 1983 estimate—14,238,000; distribution—85 per cent rural, 15 per cent urban; density—57 persons per sq. mi. (22 persons per km^2)
Largest Cities: (1975 est.) Kabul (749,000) and Qandahar (209,000)
Economy: Mainly agricultural, with some mining and limited manufacturing
Chief Products: Agriculture—barley, corn, cotton, fruits, karakul skins, nuts, rice, sugar beets, wheat, wool; manufacturing—cement, processed foods, rugs, textiles; mining—coal, gold, lapis lazuli, natural gas
Money: Basic unit—afghani
Climates: Steppe, desert, highlands

Afghanistan is a rugged land of steep mountains, rolling plains, and barren deserts. Its people are divided into various ethnic groups and tribes, many of which have their own language and customs. Nearly all Afghans are Muslims. Their religion, Islam, is the one common link among the people. Islam also has a strong influence on family life.

Afghanistan is a poor nation, and many Afghans live much as their ancestors did many years ago. There are now no railroads in Afghanistan, but plans have been made to build one linking Kabul to the rail systems of Pakistan and Iran. The country has only one radio station and one television station. Only about 10 per cent of the people can read and write. Afghanistan's farmers make a living by growing crops including wheat, cotton, fruits, and nuts. About a sixth of the people are nomads, who tend herds of livestock. Most Afghan villagers and city dwellers live in houses made of stone or sun-dried mud bricks. The nomads live in tents made of goat hair. For recreation, the people enjoy rugged outdoor games, folk songs and stories, and dancing.

Albania

Albania (2) is located on the eastern coast of the Adriatic Sea. It is bordered by Yugoslavia and Greece.

Capital: Tiranë
Official Name: Republika Popullore Socialiste e Shqipërisë, meaning People's Socialist Republic of Albania
Official Language: Albanian
Government: Communist dictatorship—ruled by Communist Albanian Workers Party, country's only political party; members carry out policies set by Politburo, party's ruling council
Flag: A red star (for Communism) above a two-headed black eagle on a red field
Area: 11,100 sq. mi. (28,748 km^2); greatest distances—north-south, 215 mi. (346 km); east-west, 90 mi. (145 km); coastline—175 mi. (282 km)
Population: 1983 estimate—2,944,000; distribution—63 per cent rural, 37 per cent urban; density—264 persons per sq. mi. (102 per km^2)
Largest Cities: (1976 est.) Tiranë (192,300), Shkodër (62,500), and Durrës (61,000)
Economy: Mainly agricultural, with some mining and forestry, but very limited manufacturing and fishing
Chief Products: Agriculture—corn, cotton, sugar beets, tobacco, wheat; mining—chromite, copper, oil; manufacturing—cement, food products, textiles
Money: Basic unit—lek
Climate: Subtropical dry summer

Albania is a small nation that has one of the lowest standards of living in Europe. Most of the people earn less than $32 a month working on government-operated farms or cooperative farms.

Albania has only about 11 cities with more than 15,000 persons. Few Albanians have cars, and there are only about 2,000 miles (3,000 kilometers) of roads that motor vehicles can use. About 65 miles (105 kilometers) of railroads link some of the larger cities.

Albania's population consists mainly of two groups—the Ghegs and the Tosks. Each group speaks its own form of the Albanian language. Turkish culture has influenced clothing styles and some other customs in Albania. Many of the men wear baggy trousers and Turkish hats called *fezzes*. Most of the women wear colorfully embroidered skirts and blouses. A majority of the people are Muslims. Most Albanians can read and write, and all children must attend elementary school for eight years.

Algeria

Algeria (3) is located in northern Africa. It is bordered by Morocco, Western Sahara, Mauritania, Mali, Niger, Tunisia, and Libya.

Capital: Algiers
Official Language: Arabic
Government: Republic—cabinet, headed by country's president, directs government operations; top members of National Liberation Front, country's only political party, have leading role in establishing policies
National Anthem: "Kassaman" ("I Take an Oath")
Flag: Left half dark green, right half white; a red star and crescent in the center
Area: 919,595 sq. mi. (2,381,741 km^2); greatest distances—north-south, 1,190 mi. (1,915 km); east-west, 1,120 mi. (1,802 km); coastline—725 mi. (1,167 km)
Population: 1983 estimate—20,377,000; distribution—61 per cent urban, 39 per cent rural; density—23 persons per sq. mi. (9 persons per km^2)
Largest Cities: (1974 est.) Algiers (1,503,720), Oran (485,139), Constantine (350,183), and Annaba (313,174)
Interesting Sights: Beautiful Muslim houses of worship called *mosques* with slender minarets, or towers from which criers called *muezzins* call people to prayer; the Museum of Antiquities in Algiers
Economy: Mainly agricultural, with some mining and limited manufacturing
Chief Products: Agriculture—fruits, grain, vegetables, wine; mining—iron ore, petroleum, phosphates
Money: Basic unit—dinar
Climates: Subtropical dry summer, steppe, desert

Algeria consists mainly of desert. The Sahara covers seven-eighths of its territory. Nearly all Algerians live and work in a narrow strip of fertile land along the country's Mediterranean coast. To keep out the heat, the people who live near the Sahara build houses with thick walls and no windows. Most Algerians herd livestock or raise crops such as wheat, citrus fruits, grapes, olives, dates, and vegetables.

Most Algerians are Arabs or Berber Muslims. Algeria was once a French territory, and its culture is a mixture of European and Arab influences. Schools offer classes in Arabic and French. Many Algerians, especially in the cities, wear traditional Arab robes. The cities have European architecture and old, crowded sections called *casbahs,* where open shops line narrow, winding streets. Algeria has airports, seaports, highways, and railroads; but camel caravans still cross the desert along routes that are thousands of years old.

Andorra

Andorra (4) is located in the Pyrenees mountains in western Europe. It is bordered by Spain and France.

Capital: Andorra
Official Name: Valls d'Andorra (Valleys of Andorra)
Official Language: Catalan
Government: Principality—bishop of Urgel, Spain, and president of France have equal powers and are jointly responsible for maintaining law and order and administering justice; must agree before changes made
Flag: National flag, used by people—blue, yellow, and red vertical stripes; state flag, used by government—same as national, but with coat of arms on yellow stripe
Area: 175 sq. mi. (453 km²); greatest distances—north-south, about 16 mi. (26 km); east-west, about 19 mi. (31 km)
Population: 1983 estimate—39,000; distribution—100 per cent rural; density—223 persons per sq. mi. (86 persons per km²)
Largest Cities: (1975 census) Andorra (10,932) and Escaldes (7,372)
Interesting Sights: Beautiful mountains; ski slopes at Pas de la Casa and Soldeu; ancient churches; country quaintness
Economy: Mainly tourism, with some farming and limited mining
Chief Products: Agriculture—tobacco; manufacturing—cigarettes and other tobacco products
Money: Basic units—French franc and Spanish peseta
Climate: Subtropical dry summer

Andorra is a tiny, mountainous country that is ruled jointly by the president of France and a bishop of Spain. Andorrans speak a language called Catalan, but most of them also understand French and Spanish. The country has two postal systems—one French and one Spanish—and two school systems. Classes are taught in French in one school system and in Spanish in the other. All Andorrans are Roman Catholics. Life for most of the people centers around their church and family.

Andorra's location, tucked among the steep Pyrenees mountains, kept it isolated from the rest of the world for hundreds of years. Most of the people were farmers or shepherds. Then, in the 1930's, roads were built that linked Andorra to France and Spain. A thriving tourist industry developed as visitors came to ski and to enjoy Andorra's beautiful scenery and charming, ancient buildings. Although some Andorrans still farm or tend livestock, many now work as hotel owners, shopkeepers or in other positions related to tourism.

Angola

Angola (5) is located on the south-western coast of Africa. It is bordered by Congo, Zaire, Zambia, and Namibia.

Capital: Luanda
Official Name: People's Republic of Angola
Official Language: Portuguese
Government: Socialist dictatorship—controlled by the Popular Movement for the Liberation of Angola (MPLA), nation's only political party; party leader serves as president, head of state, and commander in chief of armed forces; party's Central Committee, headed by premier, is chief policymaking body
Flag: Two horizontal stripes; red stripe represents Angola's struggle for independence, black stripe symbolizes Africa; yellow emblem in center has five-pointed star that stands for socialism, half cogwheel for industry, and machete for agriculture
Area: 481,354 sq. mi. (1,246,700 km^2); greatest distances—north-south, 850 mi. (1,368 km); east-west, 800 mi. (1,287 km); coastline—928 mi. (1,493 km)
Population: 1983 estimate—7,622,000; distribution—79 per cent rural, 21 per cent urban; density—16 persons per sq. mi. (6 per km^2)
Largest Cities: (1972 est.) Luanda (600,000) and Huambo (61,885)
Economy: Mainly agricultural, with limited mining and developing industries
Chief Products: Agriculture—coffee, corn, sugar cane, tobacco; manufacturing—food processing, cement, chemicals, textiles; mining—diamonds, petroleum
Money: Basic unit—kwanza
Climates: Steppe, tropical wet and dry

Angola is a large African nation that was once a colony of Portugal. Nearly all its people are black Africans who belong to various ethnic groups. There are also a small number of whites and mestizos, or persons of mixed black African and white ancestry.

Portuguese is Angola's official language, but most of the people speak one of the Bantu languages. Only about 15 per cent of all Angolans can read and write. About half the people are Christians. Most of the others practice religions based on worship of ancestors and spirits.

About 80 per cent of all Angolans live in rural areas. They work as farmers and herders, and many of them raise just enough food for their own use. Petroleum is Angola's leading export, and the country also has huge deposits of diamonds, iron ore, and copper.

Antigua and Barbuda

Antigua and Barbuda (6) is an island country in the Caribbean Sea about 430 miles (692 kilometers) north of Venezuela. It is made up of three islands—Antigua, Barbuda, and Redonda.

Capital: St. John's
Official Language: English
Government: Constitutional monarchy—government operations conducted by prime minister and Cabinet; two-house, elected Parliament makes country's laws
Flag: Bottom corners, red; center triangular portion, black over blue over white with a yellow half sun superimposed on black area
Area: Total—171 sq. mi. (442 km^2); Antigua—108 sq. mi. (280 km^2), Barbuda—62 sq. mi. (161 km^2), Redonda—½ sq. mi. (1.3 km^2)
Population: 1983 estimate—79,000; distribution—66 per cent rural, 34 per cent urban; Antigua—98 per cent of population, Barbuda—2 per cent of population, Redonda—uninhabited; density—482 persons per sq. mi. (186 persons per km^2)
Largest City: (1975 est.) St. John's (24,000)
Economy: Mainly tourism, with some farming and manufacturing
Chief Products: Agriculture—sugar cane, cotton; manufacturing—clothing, paint, large appliances
Money: Basic unit—East Caribbean dollar
Climate: Tropical wet

Antigua and Barbuda is a nation of three islands formed by volcanoes. The islands of Antigua and Barbuda are mostly flat, having been worn down by wind and rain. Redonda is rocky with little plant life and no human inhabitants. The islands are a haven for tourists who are lured by the country's many resorts, its beautiful white, sandy beaches, and its warm, sunny climate. Drought, however, does strike the area for long periods from time to time.

Most of the people of Antigua and Barbuda are descendants of black Africans. About a third of the country's population resides in St. John's, located on the northwest coast of Antigua. English is spoken almost exclusively and the country's primary and secondary educational systems are well-developed.

Farmers in Antigua and Barbuda raise sugar cane, which is refined into sugar by processors in the country, and cotton. The government has encouraged the development of small industries to bolster the economy.

Argentina

Argentina (7) is located in south-eastern South America. It is bordered by Chile, Bolivia, Paraguay, Brazil, and Uruguay.

Capital: Buenos Aires
Official Name: República Argentina (Republic of Argentina)
Official Language: Spanish
Government: Military rule (officially a republic)—a three-member military council called a *junta* makes government policies
National Anthem: "Himno Nacional Argentino" ("Argentine National Anthem")
Flag: One white horizontal stripe between two blue stripes; "Sun of May" in center of white stripe symbolizes Argentina's freedom from Spain
Area: 1,072,163 sq. mi. (2,776,889 km^2); greatest distances—north-south, 2,300 mi. (3,701 km); east-west, 980 mi. (1,577 km); coastline—2,940 mi. (4,731 km)
Population: 1983 estimate—28,964,000; distribution—82 per cent urban, 18 per cent rural; density—26 persons per sq. mi. (10 persons per km^2)
Largest Cities: (1980 census) Buenos Aires (2,908,001), Córdoba (968,664), and Rosario (875,623)
Economy: Mainly agriculture, industrial production, service industries, and trade
Chief Products: Agriculture—alfalfa, cattle, citrus fruits, corn, cotton, flaxseed, maté, rye, sheep, wheat; manufacturing and processing—beef, hides, textiles, wool; mining—building stone, coal, iron, petroleum, salt
Money: Basic unit—peso
Climates: Varied—oceanic moist, subtropical moist, steppe, desert, highlands, tropical wet and dry

Argentina is one of the world's leading agricultural nations. The rich soil and grazing land of a region called the Pampa produce livestock and grain that are the source of much of the country's wealth. Argentina leads the world in beef exports and is one of the chief producers of sheep, wool, and wheat. Argentine cowhands, called *gauchos,* tend huge herds of cattle in the Pampa.

Most Argentines are descendants of people who immigrated from Spain and Italy. They have one of the high standards of living in Latin America. About 90 per cent of all Argentines can read and write.

Life in Argentina shows a European influence. Spanish-style buildings are common in many towns. Most of the people take a siesta or rest period after lunch and eat dinner late at night. Most Argentines are Roman Catholics.

Australia

Australia (8) is both a continent and a country. It lies between the Indian Ocean and the Pacific Ocean.

Capital: Canberra
Official Name: Commonwealth of Australia
Official Language: English
Government: Constitutional monarchy—two-house, parliamentary system of government headed by prime minister who appoints Cabinet; British monarch, Queen Elizabeth II, acts as nation's head of state but has no power in government
Anthems: "Advance Australia Fair" (national); "God Save the Queen" (royal)
Flag: British Union Flag in top left corner, five white stars for the constellation Southern Cross, and a large white star for the Commonwealth, all on a field of blue
Area: 2,966,150 sq. mi. (7,682,300 km²); greatest distances (mainland)—east-west, 2,475 mi. (3,983 km); north-south, 1,950 mi. (3,138 km); coastline—17,366 mi. (27,948 km)
Population: 1983 estimate—15,150,000; distribution—89 per cent urban, 11 per cent rural; density—5 persons per sq. mi. (2 persons per km²)
Largest Cities: (1981 census) Sydney (2,874,415), Melbourne (2,578,527), Brisbane (942,636), and Perth (809,033)
Economy: Mainly agriculture and mining, with increasing manufacturing
Chief Products: Agriculture—wool, wheat, cattle and calves, dairy products, sugar cane, fruits, sheep and lambs, barley; manufacturing—processed foods, iron, steel, other metals, transportation equipment, paper, household appliances; mining—bauxite, coal, copper, gold, iron ore, lead, manganese, natural gas, nickel, opals, petroleum, silver, tin, tungsten, uranium, zinc
Money: Basic unit—dollar
Climates: Oceanic moist, subtropical moist, subtropical dry summer, steppe, desert, tropical wet and dry

Australia is a large, thriving nation that produces a variety of agricultural, mineral, and manufactured products. Nearly all Australians are of European ancestry, mainly British and Irish. Aborigines, a dark-skinned people who were Australia's original inhabitants, make up less than 1 per cent of the population.

About 89 per cent of Australia's people live in cities and towns. The cities have high-rise office buildings, fine shops, theaters, and restaurants. Unlike American cities, those in Australia have few apartment buildings. Most families live in one-story houses with a yard and garden. Australians love outdoor sports, and the country's warm, pleasant climate enables them to take part in all kinds of outdoor recreation year-round.

Austria

Austria (9) is located in central Europe. It is bordered by Switzerland, Liechtenstein, West Germany, Czechoslovakia, Hungary, Italy, and Yugoslavia.

Capital: Vienna
Official Name: Republik Österreich
(Republic of Austria)
Official Language: German
Government: Federal republic—elected president serves as head of state, duties largely ceremonial; government run by chancellor and Cabinet
National Anthem: "Land der Berge, Land am Strome" ("Land of Mountains, Land at the River")
Flag: White horizontal stripe between two red stripes with coat of arms in center
Area: 32,374 sq. mi. (83,849 km^2); greatest distances—east-west, 355 mi. (571 km); north-south, 180 mi. (290 km)
Population: 1983 estimate—7,530,000; distribution—54 per cent urban, 46 per cent rural; density—233 persons per sq. mi. (90 persons per km^2)
Largest Cities: (1971 census) Vienna (1,614,841), Graz (248,500), Linz (202,874), Salzburg (128,845), and Innsbruck (115,197)
Economy: Mainly manufacturing and trade, with some agriculture and forestry, but limited mining
Chief Products: Agriculture—barley, corn, dairy products, livestock, oats, potatoes, rye, sugar beets, wheat; manufacturing—cement, chemical products, electrical equipment, furniture, glass, iron and steel, leather goods, lumber, machines and tools, motor vehicles, optical instruments, paper and pulp, processed foods and beverages, textiles and clothing; mining—coal, copper, graphite, iron ore, lead, magnesite, natural gas, oil, salt, zinc
Money: Basic unit—schilling
Climates: Continental moist, highlands

Austria is a prosperous nation known for its beautiful mountain scenery and long cultural tradition. Most Austrians live in cities and towns, and nearly a third of them work in manufacturing industries. About 90 per cent of all Austrians are Roman Catholics.

Austrians love good food and beverages. Coffee and delicious pastries are a favorite snack. Holidays and festivals are an important part of life in Austria, and many people dress up in traditional costumes for special occasions.

Austrians are proud of their country's contributions to the arts, particularly music. Some of the world's great composers of classical music lived in Austria.

Bahamas

Bahamas (10) is an island nation in the northern part of the West Indies.

Capital: Nassau
Official Name: The Commonwealth of the Bahamas
Official Language: English
Government: Constitutional monarchy—two-house, parliamentary legislature; head of party holding most parliamentary seats serves as prime minister; British monarch, Queen Elizabeth II, is official head of state; governor general represents her in Bahamas
National Anthem: "March On, Bahamaland"
Flag: Black triangle represents the Bahamian people; blue and gold horizontal stripes stand for sea and land
Area: 5,380 sq. mi. (13,935 km^2); greatest distances—north-south, 450 mi. (724 km); east-west, 435 mi. (700 km); coastline—1,580 mi. (2,543 km)
Population: 1983 estimate—240,000; distribution—58 per cent urban, 42 per cent rural; density—44 persons per sq. mi. (17 persons per km^2)
Largest Cities: (1970 census) Freeport (15,277) and Nassau (3,233)
Economy: Mainly tourism and related activities
Chief Products: Agriculture—bananas, citrus fruits, cucumbers, pineapples, tomatoes; manufacturing—cement, food products, petroleum products, rum
Money: Basic unit—Bahamian dollar
Climate: Tropical wet and dry

Bahamas has a mild climate, fine beaches, and tropical beauty which attract more than a million tourists a year. The nation consists of about 3,000 coral islands and reefs, but people live on only about 20 of the islands. Roughly three-fourths of all Bahamians live on two of the islands: New Providence and Grand Bahama.

About 80 per cent of the people of the Bahamas are blacks. Most of the others are whites and mulattoes or persons of mixed black and white ancestry. Bahamian children from the ages of 5 to 14 must attend school, and more than 90 per cent of the people can read and write.

Many Bahamians work in hotels, restaurants, and other businesses related to tourism. Less than 2 per cent of the country's workers are farmers because most of the islands are covered with stony, infertile soil. The farmers grow bananas, citrus fruits, cucumbers, pineapples, tomatoes, and other crops. Bahamian fishermen catch a variety of seafood, but the nation must import most of its food.

Bahrain

Bahrain (11) is located in the Persian Gulf in Southwest Asia.

Capital: Manama
Official Name: The State of Bahrain
Official Language: Arabic
Government: Emirate—headed by ruler called an *emir* who appoints a twelve-member Council of Ministers that runs the government
National Anthem: "Assalam al-Amiri" ("National Anthem")
Flag: Red field covers about three-fourths of flag and adjoins jagged edge of a vertical white stripe
Area: 240 sq. mi. (622 km^2); greatest distances—north-south, 50 mi. (80 km); east-west, 26 mi. (42 km); coastline—78 mi. (126 km)
Population: 1983 estimate—414,000; distribution—78 per cent urban, 22 per cent rural; density—1,725 persons per sq. mi. (666 persons per km^2)
Largest Cities: (1975 est.) Manama (94,697) and Al Muharraq (44,567)
Economy: Mainly the production of crude oil and petroleum refining with limited construction, fishing, and manufacturing
Chief Products: Agriculture—dairy cattle, dates and other fruits, grains, vegetables, poultry; fishing—shrimp; manufacturing—aluminum, boats, building materials, petroleum products, plastics products, soft drinks; mining—petroleum
Money: Basic unit—dinar; one thousand fils equal one dinar
Climate: Desert

Bahrain is a tiny island nation with one of the high standards of living in the Persian Gulf area. The oil industry has made Bahrain prosperous. Bahrain has only a small oil supply of its own, but it has a large, modern refinery that processes all of its oil as well as much that comes by pipeline from Saudi Arabia.

About 90 per cent of all Bahrainis are Arabs, and Islam is the national religion. Minority groups include Indians, Iranians, and Pakistanis. Arabic is the official language, but some people speak English and Persian as well. The government provides free education and medical care. Bahrain has over 100 elementary and high schools.

Most of Bahrain is desert. The majority of its people live in cities and villages in the northern part of the island of Bahrain. This area also has irrigated farmland where farmers grow grains, dates and other fruits, and vegetables. An excellent electric supply system makes conveniences such as air conditioners and refrigerators common in both urban and rural areas of Bahrain.

Bangladesh

Bangladesh (12) is located in South Asia. It is bordered by India and Burma.

Capital: Dacca
Official Name: People's Republic of Bangladesh
Also Called: East Bengal
Official Language: Bengali
Government: Military rule—military leaders took control in 1982 and rule under martial law
National Anthem: "Amar Sonar Bangla" ("My Golden Bengal")
Flag: Large red circle symbolizing the sun on a dark green field which stands for scenic beauty
Area: 55,598 sq. mi. (143,998 km^2); greatest distances—north-south, 464 mi. (747 km); east-west, 288 mi. (463 km); coastline—357 mi. (575 km)
Population: 1983 estimate—91,281,000; distribution—89 per cent rural, 11 per cent urban; density—1,642 persons per sq. mi. (634 persons per km^2)
Largest Cities: (1974 census) Dacca (1,679,572), Chittagong (889,760), and Khulna (437,304)
Economy: Mainly agricultural, with limited industry
Chief Products: Agriculture—hides and skins, jute, rice, sugar cane, tea, tobacco; manufacturing—jute products, paper and paper products, textiles; mining—natural gas
Money: Basic unit—taka
Climates: Tropical wet and dry, tropical wet

Bangladesh is a small, densely populated country. Overcrowding is a serious problem even in rural areas, where most of the people live. The majority of Bangalees are poor farmers who use outdated tools and methods to make a bare living from the land. Rice and fish are the chief foods. Many of the people do not have enough to eat. A typical village home in Bangladesh consists of one or two rooms with no plumbing or electricity. Small wooden houses crowd the cities, and many urban people live in slums.

More than 95 per cent of the people of Bangladesh are Bengalis, a dark-skinned people who speak a language also called Bengali. About 85 per cent are Muslims, and most of the rest are Hindus.

Bangladesh has a warm, humid climate, and people throughout the country wear loose, lightweight clothing. Countless rivers and streams crisscross the land, providing the country's chief transportation routes. Boats can travel to almost every part of Bangladesh.

Barbados

Barbados (13) is located in the West Indies. It lies about 250 miles northeast of the South American mainland.

Capital: Bridgetown
Official Language: English
Government: Constitutional monarchy—governor general represents British Crown as head of state; prime minister and Cabinet actually govern country; two-house Parliament consisting of Senate and House of Assembly
Flag: Two blue outer stripes represent sea and sky; center gold stripe stands for sandy beaches; black trident head with broken shaft in center of flag symbolizes Neptune, the sea god, and the change from dependence to independence
Area: 166 sq. mi. (431 km^2); greatest distances—north-south, 21 mi. (34 km); east-west, 14 mi. (23 km); coastline—56 mi. (90 km)
Population: 1983 estimate—253,000; distribution—61 per cent rural, 39 per cent urban; density—1,520 persons per sq. mi. (587 persons per km^2)
Largest City: (1972 est.) Bridgetown (8,782)
Economy: Mainly tourism, commerce and finance, and agriculture
Chief Products: Agriculture—sugar; fishing—flying fish, shrimp; manufacturing—molasses, rum
Money: Basic unit—Barbados dollar
Climate: Tropical wet

Barbados is a tiny island nation where sugar growing and tourism employ nearly half the work force. Sugar cane grows on about 70 per cent of the farmland of Barbados. About a fourth of the country's workers labor on large sugar plantations or on their own small farms, where they grow various food crops. Its warm climate and sandy beaches have led to the development of a tourist industry that employs many other workers.

Barbados was a British colony for more than 300 years. Most of its people are blacks, descended from African slaves brought to the island several hundred years ago. Over 15 per cent are of mixed black and British ancestry, and most of the rest are of British descent.

British influence is evident in several aspects of life in Barbados. All of the people speak English, and the most popular sport is cricket. Members of the Church of England rank as the largest religious group. About 90 per cent of the people can read and write.

Belgium

Belgium (14) is located in northwestern Europe. It is bordered by The Netherlands, West Germany, Luxembourg, and France.

Capital: Brussels
Official Name: Kingdom of Belgium
Official Languages: Dutch and French
Government: Constitutional monarchy—
    two-house Parliament; king serves as head of
    state but with little real power; executive power
    lies in hands of prime minister and members of Council of Ministers
    or cabinet; cabinet must consist of equal numbers of Dutch speakers and
    French speakers
National Anthem: "La Brabançonne" ("The Brabant Song")
Flag: Three vertical stripes of black, yellow, and red
Area: 11,781 sq. mi. (30,513 km^2); greatest distances—east-west, 170 mi. (274 km); north-south, 140 mi. (225 km); coastline—39 mi. (63 km)
Population: 1983 estimate—9,887,000; distribution—95 per cent urban, 5 per cent rural; density—839 persons per sq. mi. (324 persons per km^2)
Largest Cities: (1975 est.—metro area) Brussels (1,050,787), Antwerp (662,317), and Ghent (218,526)
Economy: Mainly industrial and commercial, with limited agriculture
Chief Products: Agriculture—barley, cattle, dairy products, flax, hops, oats, potatoes, sugar beets, wheat; manufacturing—cement, chemicals and chemical products, glass, leather goods, paper, steel, textiles
Money: Basic unit—Belgian franc
Climate: Oceanic moist

Belgium is a small nation populated by two ethnic groups: Flemings and Walloons. The Flemings, who live mainly in northern Belgium, speak Dutch. The Walloons, who live mainly in the south, speak French. The difference in their cultures has long been a source of friction between the two groups. Members of both groups live in Brussels, the nation's capital.

About 98 per cent of all Belgians are Roman Catholic. Religion is particularly important in the education and social life of the Flemings. Belgians in general are well educated and enjoy a high standard of living. Bicycle riding and soccer rank as leading sports in Belgium.

About 95 per cent of all Belgians live in cities and towns. Manufacturing industries employ nearly a third of the nation's workers. Only about 5 per cent are farmers, but they produce nearly all the food Belgium needs. Steel production is Belgium's chief industry, but the country is well-known for its textiles, especially delicate Belgian lace.

Belize

Belize (15) is located in Central
America on the southeast coast of
Yucatán Peninsula. It is bordered
by Guatemala, Mexico, and the Caribbean Sea.

Capital: Belmopan
Formerly Called: British Honduras
Official Language: English
Government: Constitutional monarchy—functions
as parliamentary democracy; prime minister heads
government aided by 12 Cabinet ministers; two-house National Assembly
makes country's laws
National Anthem: "Land of the Free"
Flag: A wide horizontal blue stripe bordered by narrow horizontal red stripes at
top and bottom; country's coat of arms appears on blue stripe
Area: 8,867 sq. mi. (22,965 km^2); greatest distances—north-south, 180 mi. (290
km); east-west, 85 mi. (137 km); coastline—220 mi. (354 km)
Population: 1983 estimate—158,000; distribution—62 per cent urban, 38 per
cent rural; density—18 persons per sq. mi. (7 persons per km^2)
Largest City: (1972 est.) Belize City (41,500)
Economy: Mainly agricultural, but with limited fishing and industry
Chief Products: Agriculture and forestry—sugar cane, bananas, grapefruit, or-
anges, cedar, mahogany; fishing industry—crabs, crayfish, groupers; manu-
facturing and processing—clothing, construction materials
Money: Basic unit—Belizean dollar
Climates: Tropical wet and dry, tropical wet

Belize is a tiny agricultural nation that ranks as Central America's most
thinly populated country. Over 50 per cent of its people live along the
Caribbean coast. From the mid-1800's until 1973, Belize was ruled by
Great Britain and was called British Honduras. It changed its name in
1973 and gained independence in 1981.

Belize has an extremely mixed racial population. About 50 per cent of
the people have full or part black African ancestry. About 20 per cent of
the people are descended from Maya or other Indian groups and nearly
20 per cent are mestizos, or persons of mixed European and Indian ancestry. Most of the rest are of European, East Indian, Chinese, or Leb-
anese descent. Although English is the nation's official language, many
people also speak Spanish and some speak Maya Indian languages. Roman Catholic and Protestant are the two major religious groups. Unemployment in urban areas and poor farm production account for the low
standard of living among Belizeans.

Benin

Benin (16) is located in West Africa. It is bordered by Upper Volta, Niger, Nigeria, and Togo.

Capital: Porto-Novo
Official Name: République Populaire du Benin (Popular Republic of Benin)
Official Language: French
Government: Republic—Council of Ministers, headed by president, directs government; president and other council members are military officers
National Anthem: "L'Aube Nouvelle" ("The New Dawn")
Flag: Green with a red star in upper left corner; green stands for agriculture, the basis of the economy; star symbolizes the socialist government
Area: 43,484 sq. mi. (112,622 km^2); greatest distances—north-south, 415 mi. (668 km); east-west, 202 mi. (325 km); coastline—77 mi. (124 km)
Population: 1983 estimate—3,875,000; distribution—69 per cent rural, 31 per cent urban; density—88 persons per sq. mi. (34 persons per km^2)
Largest Cities: (1975 est.) Cotonou (178,000) and Porto-Novo (104,000)
Economy: Mainly agricultural
Chief Products: Agriculture—coffee, cotton, palm oil and kernels, peanuts, shea nuts, tobacco
Money: Basic unit—franc
Climates: Steppe, tropical wet and dry, tropical wet

Benin is a small agricultural nation that depends largely on palm trees as its source of wealth. Its most important exports are palm oil and palm kernels, which are used to make soap and margarine.

Most of Benin's people are black, and they are divided into about 60 different ethnic groups. Benin was once a French territory, and French is still the official language. More than 30 per cent of Benin's children attend school. About two-thirds of the people practice animism, or the belief that all objects in nature have spirits. Most of the rest of the people are Christians or Muslims.

Farming is the chief occupation in Benin. In addition to various export crops, the people raise livestock and grow foods such as beans, corn, manioc, rice, and sweet potatoes. In the lagoon areas near Benin's coast, the people live in simple bamboo houses built on poles to keep them out of the water. The people in the mountain areas of the northwest live in round houses with mud walls and thatched roofs.

Bhutan

Bhutan (17) is located in south-central Asia. It is bordered by China and India.

Capital: Thimphu
Official Language: Druk-ke, a Tibetan dialect
Government: Monarchy—a hereditary, or inherited, monarchy headed by a powerful king who appoints prime minister and an advisory council to assist him; king also appoints one-fourth of the 130-member national assembly called the *Tsongdu*
Flag: Square flag divided diagonally into yellow and orange halves; white dragon in center has a jewel in each claw
Area: 18,147 sq. mi. (47,000 km^2); greatest distances—north-south, 110 mi. (177 km); east-west, 200 mi. (322 km)
Population: 1983 estimate—1,390,000; distribution—96 per cent rural, 4 per cent urban; density—77 persons per sq. mi. (30 persons per km^2)
Largest City: (1977 est.) Thimphu (8,922)
Economy: Almost entirely agricultural and pastoral, but with very limited mining
Chief Products: Agriculture—barley, fruit, rice, vegetables, wheat; handicrafts and industries—blankets, leatherwork, pottery, preserved fruit, textiles; mining—coal
Money: Basic unit—Indian rupee
Climate: Highlands

Bhutan is a rugged land in the Himalayas. The mountains cut Bhutan off from the rest of the world until about 1960, when, with the help of India, Bhutan began to build roads and power stations. Most Bhutanese still have little contact with anyone outside their own village. About 95 per cent of the people cannot read or write.

Most Bhutanese live in isolated mountain valleys. The people grow wheat, barley, rice, fruits, and vegetables on irrigated terraces built into the mountain slopes. In the higher mountain regions, the people tend herds of cattle and yaks. Many Bhutanese live in houses made of mud blocks and stone. The living quarters are above a ground floor that is used as a barn. The people who live in the low, forested foothills build their houses on high ground to protect themselves against floods, wild animals, and snakes. Bhutan's official religion is a form of Buddhism called Lamaism. A minority of the people are Hindus.

Bolivia

Bolivia (18) is located in South America. It is bordered by Brazil, Paraguay, Argentina, Chile, and Peru.

Capitals: Sucre (official); La Paz (actual)
Official Name: República de Bolivia (Republic of Bolivia)
Official Language: Spanish
Government: Republic—elected president heads government; military leaders have often taken control of government
National Anthem: "Himno Nacional de Bolivia" ("National Anthem of Bolivia")
Flag: The national flag, flown by the people—horizontal red, yellow, and green stripes (top to bottom); state flag, used by the government—same flag with Bolivia's coat of arms at the center
Area: 424,164 sq. mi. (1,098,581 km²); greatest distances—north-south, 900 mi. (1,448 km); east-west, 800 mi. (1,287 km)
Population: 1983 estimate—6,048,000; distribution—58 per cent rural, 42 per cent urban; density—16 persons per sq. mi. (6 persons per km²)
Largest Cities: (1976 census) La Paz (654,713) and Santa Cruz (256,946)
Economy: Mainly agricultural, with growing manufacturing and mining
Chief Products: Agriculture—cattle, coca, coffee, corn, cotton, rice, hides, sheep, sugar; forestry—rubber, quebracho; manufacturing and processing—beverages, processed foods, textiles; mining—antimony, bismuth, copper, gold, lead, oil, tin, tungsten, silver, zinc
Money: Basic unit—peso boliviano
Climates: Steppe, highlands, tropical wet and dry, tropical wet

Bolivia has many mineral resources. Mining is the most important industry, though only about 2 per cent of the country's workers are miners. Tin and other metals account for nearly all of Bolivia's exports.

More than half of all Bolivians are Indians who make a bare living as farmers in the country's dry, western plateau region. They use primitive tools and methods to raise sheep and pigs and to grow their food, which includes corn, potatoes, and a cereal called *quinoa*. Llamas, alpacas, and vicuñas provide wool that is made into cloth.

About 42 per cent of Bolivia's people live in cities and towns. Most of the urban dwellers are whites and mestizos, or persons of mixed white and Indian ancestry. The whites and most of the mestizos speak Spanish, Bolivia's official language. The Roman Catholic Church is the state church of Bolivia. Religious holidays and festivals are an important part of Bolivian life. Soccer is the nation's most popular sport.

Botswana

Botswana (19) is located in southern Africa. It is bordered by Namibia, Zimbabwe, and South Africa.

Capital: Gaborone
Official Name: Republic of Botswana
Official Language: English
Government: Republic—headed by president who is elected by the National Assembly, country's chief legislative body; president selects a Cabinet from National Assembly; leaders of nation's major ethnic groups make up House of Chiefs and advise government on matters affecting ethnic customs
National Anthem: "Fatshe La Rona" ("Blessed Country")
Flag: Three horizontal bands (blue, black, and blue) are divided by two narrow white bands
Area: 231,805 sq. mi. (600,372 km²); greatest distances—north-south, 625 mi. (1,006 km); east-west, 590 mi. (950 km)
Population: 1983 estimate—885,000; distribution—70 per cent rural, 30 per cent urban; density—3 persons per sq. mi. (1 person per km²)
Largest Cities: (1971 census) Gaborone (18,799), Francistown (18,613), and Serowe (15,723)
Economy: Mainly agriculture and herding, but with a growing mining industry
Chief Products: Agriculture—beans, cattle, corn, cowpeas, goats, hides and skins, meat, millet, sorghum; mining—copper, diamonds, nickel
Money: Basic unit—pula
Climates: Steppe, desert, tropical wet and dry

Botswana is a thinly populated nation covered largely by the Kalahari Desert. Most Botswanans live in the eastern part of the country, where the land is somewhat fertile. A large majority are black Africans called Tswana. Most of them live in large rural villages and farm or raise livestock. Botswana also has about 10,000 Bushmen who follow a primitive way of life in the desert, gathering food and hunting much as their ancestors did thousands of years ago. Several thousand whites also live in Botswana.

Farmers of Botswana grow corn, millet, and sorghum, among other crops. Cattle and goats provide meat, hides, and skins, which are important exports. Botswana's large deposits of diamonds, copper, nickel, and other minerals are being developed. About 40,000 Botswanans work in neighboring South Africa for part of the year because there is not enough work in their own country. Most of Botswana's roads are unpaved, and its exports and imports are transported via a railroad that runs through South Africa.

Brazil

Brazil (20) is located in South America. It is bordered by every country in South America except Chile and Ecuador.

Capital: Brasília
Official Name: República Federativa do Brasil (Federative Republic of Brazil)
Official Language: Portuguese
Government: Federal republic—headed by powerful president; military leaders play important role in country's government
National Anthem: "Hino Nacional"
Flag: A blue sphere speckled with white stars centered in yellow diamond shape which is centered on a field of green; sphere bears the motto "Order and Progress"; green and yellow symbolize Brazil's forests and minerals
Area: 3,286,487 sq. mi. (8,511,965 km²); greatest distances—north-south, 2,684 mi. (4,319 km); east-west, 2,689 mi. (4,328 km); coastline—6,019 mi. (9,687 km)
Population: 1983 estimate—127,427,000; distribution—68 per cent urban, 32 per cent rural; density—39 persons per sq. mi. (15 persons per km²)
Largest Cities: (1980 census) São Paulo (7,003,529), Rio de Janeiro (5,093,232), Belo Horizonte (1,442,483), and Recife (1,184,215)
Economy: Mainly industry and services
Chief Products: Agriculture—bananas, cacao beans, cattle, coffee, corn, oranges, soybeans, sugar cane; manufacturing and processing—automobiles, cement, chemicals, electrical equipment, food products, paper, railroad cars, rubber, steel, textiles, trucks; mining—beryllium, chrome, diamonds, iron ore, magnesite, manganese, mica, quartz crystals, tin, titanium; forest products—Brazil nuts, carnauba wax, latex, timber
Money: Basic unit—cruzeiro
Climates: Subtropical moist, steppe, tropical wet and dry, tropical wet

Brazil is a huge nation, rich in natural resources. Vast forests, large mineral deposits, and water power provide raw materials for various industries. About two-thirds of the people live in urban areas. Brazil's farms produce almost all the country's food and a variety of export crops including coffee. Agriculture employs nearly a third of all Brazilian workers, but few farmers own their own land. Most are paid laborers.

Many Brazilians are of European descent. Others are black or are of mixed ancestry. Only about 1 per cent are Indian. Portuguese is Brazil's official language.

Standards of living vary widely in Brazil. Housing ranges from high-rise apartment buildings in the cities to farmhouses made of branches plastered with mud. Beans and rice are common foods.

Bulgaria

Bulgaria (21) is located in south-eastern Europe on the Balkan Peninsula. It is bordered by Romania, Turkey, Greece, and Yugoslavia.

Capital: Sofia
Official Name: Narodna Republika Bulgariya (People's Republic of Bulgaria)
Official Language: Bulgarian
Government: People's Republic (Communist dictatorship)— Bulgarian Communist Party controls government; power centered in Politburo, part of Communist Party's central committee; one-house legislature, the National Assembly, makes the country's laws
National Anthem: "Mila Rodino" ("Dear Fatherland")
Flag: Horizontal white, green, and red stripes (top to bottom); coat of arms at upper left has a lion surrounded by wheat and topped by a red star
Area: 42,823 sq. mi. (110,912 km^2); greatest distances—east-west, 306 mi. (492 km); north-south, 170 mi. (274 km); coastline—175 mi. (282 km)
Population: 1983 estimate—9,049,000; distribution—64 per cent urban, 36 per cent rural; density—212 persons per sq. mi. (82 persons per km^2)
Largest Cities: (1975 census) Sófia (965,729) and Plovdiv (300,242)
Economy: Primarily industrial, with some agriculture
Chief Products: Agriculture—corn, grapes and other fruit, poultry, roses, sheep, tobacco, vegetables, wheat; manufacturing and processing—chemicals, flour, machinery, metallurgic products, rose oil; mining—bauxite, coal, copper, lead, uranium, zinc
Money: Basic unit—lev; one hundred stotinki equal one lev
Climates: Continental moist, steppe

Bulgaria is a nation whose people are mainly of mixed European and Asian ancestry. Most Bulgarians are a blend of two ethnic groups, the European Slavs and the Asian Bulgars, who lived more than a thousand years ago in what is now Bulgaria. Present-day Bulgarians speak a Slavic language also called Bulgarian.

The government owns all of Bulgaria's industries and nearly all its farmland. Farmers work on large collective farms that produce barley, corn, rye, wheat, and tobacco, among other crops. Damask roses grown in central Bulgaria are used in making perfume.

More than 60 per cent of all Bulgarians live in urban areas, and many work in manufacturing industries. Most city dwellers live in apartment buildings. Many rural houses are modern brick structures with electricity and running water. The government provides free education, medical care, and vacation resorts.

Burma

Burma (22) is located in Southeast Asia. It is bordered by China, Laos, Thailand, Bangladesh, and India.

Capital: Rangoon
Official Name: Pyi-Daung-Su-Shay-Lis-Tha-Ma-Da-Myan-Ma-Naing-Ngan-Daw (The Socialist Republic of the Union of Burma)
Official Language: Burmese
Government: Republic—one-house legislature, called People's Assembly, elects a State Council from among members whose chairman serves as president
Anthem: "A-Myo-tha The-chin" ("National Anthem")
Flag: Blue rectangle at top left on red field; fourteen white stars surround a cogwheel and a rice plant; the stars stand for the 14 states, and the cogwheel and rice for industry and agriculture
Area: 261,218 sq. mi. (676,552 km²); greatest distances—north-south, 1,200 mi. (1,931 km); east-west, 625 mi. (1,006 km); coastline—1,650 mi. (2,655 km)
Population: 1983 estimate—37,670,000; distribution—78 per cent rural, 22 per cent urban; density—145 persons per sq. mi. (56 persons per km²)
Largest Cities: (1979 est.) Rangoon (1,315,964) and Mandalay (472,512)
Economy: Mainly agriculture and forestry, with some limited mining
Chief Products: Agriculture—corn, cotton, peanuts, rice, rubber, sesame seed, sugar, tea, tobacco; mining—antimony, copper, gold, lead, petroleum, tin, tungsten, zinc; forest products—bamboo, ironwood, teak; manufacturing and processing—handicraft articles, silk
Money: Basic unit—kyat
Climates: Highlands, tropical wet and dry, tropical wet

Burma ranks as one of the world's leading rice-growing nations. The hot, wet climate and fertile soil along the Irrawaddy River are ideal for growing rice, and over half of Burma's farmland is used for this purpose.

About four-fifths of the Burmese people live in rural areas. Most of them are farmers who live on the food they grow. Others work in forests where teak, bamboo, and other valuable trees grow. Most Burmese live in thatched bamboo houses set on stilts to protect against wild animals and insects. An open porch serves as a living and dining room. Rice is the chief food, and the people usually eat it with their fingers.

About 85 per cent of all Burmese are Buddhists. Buddhist monasteries, temples, and shrines, called *pagodas,* are a characteristic feature of Burmese villages and cities. Religious festivals provide a major form of recreation. Many festivals feature actors, dancers, and singers in open-air theatrical performances.

Burundi

Burundi (23) is located in central Africa. It is bordered by Rwanda, Tanzania, and Zaire.

Capital: Bujumbura
Official Languages: Kirundi and French
Government: Republic—in practice, country governed by group of military leaders called the Supreme Military Council headed by a president
Flag: From a white circle in center, white bands extend to the corners; the field is red above and below the circle, and green to the left and right of it; in the circle are three red stars rimmed in green
Area: 10,747 sq. mi. (27,834 km²); greatest distances—north-south, 150 mi. (241 km); east-west, 135 mi. (217 km)
Population: 1983 estimate—4,355,000; distribution—98 per cent rural, 2 per cent urban; density—404 persons per sq. mi. (156 persons per km²)
Largest Cities: (1976 est.) Bujumbura (157,000); (1970 est.) Muyinga (19,000)
Economy: Mainly agricultural, with very limited industry
Chief Products: Agriculture—bananas, barley, beans, coffee, corn, cotton, livestock, manioc, palm oil, peanuts, peas, rice, sorghum, squash, sweet potatoes, tea, tobacco, wheat, yams; mining—tin; fishing—freshwater fish
Money: Basic unit—franc
Climate: Tropical wet and dry

Burundi is a tiny, underdeveloped nation with few natural resources. It has dirt roads and trails, but no railroads. Its inland location makes overseas trade difficult and costly. However, the capital city of Bujumbura does have an international airport.

Almost all of Burundi's people live in rural areas. About 84 per cent of them belong to the Bahutu ethnic group. They are farmers who grow just enough beans, corn, manioc, and sweet potatoes to feed their families. Coffee is the most important crop grown for export. About 15 per cent of the people are Watusi who raise cattle for almost all their food and other needs. Burundi's population also includes Pygmies. Many of these make pottery for a living.

Burundi has two official languages: French and Kirundi, a Bantu language spoken by most of the people. Nearly half the people are Roman Catholics. Most of the rest practice religions that include a belief in the magical powers of various objects. Christian missionaries operate most of the schools.

Cambodia

Cambodia (24) is located in Southeast Asia. It is bordered by Thailand, Laos, and Vietnam.

Capital: Phnom Penh
Official Name: Democratic Kampuchea
Official Language: Cambodian (Khmer)
Government: Communist dictatorship—
   People's Revolutionary Council heads
   government; Council's president is top government official
Flag: Yellow design of the five-towered temple of Angkor Wat in the center of a red field; the temple is the national symbol of the country; the red field symbolizes Communism
Area: 69,898 sq. mi. (181,035 km²); greatest distances—east-west, 350 mi. (563 km); north-south, 280 mi. (451 km); coastline—220 mi. (354 km)
Population: 1983 estimate—9,360,000, based on a United Nations estimate; but many experts estimate the population at about 5,000,000; distribution—86 per cent rural, 14 per cent urban; density—135 persons per sq. mi. (52 persons per km²)
Largest Cities: Data unavailable
Economy: Mainly agricultural
Chief Products: Agriculture—cattle, rice, rubber, soybeans; manufacturing and processing—cement, paper, plywood, processed rice and fish, textiles
Climates: Tropical wet and dry, tropical wet

Cambodia is an agricultural nation covered by fertile flood plains and forests. The country has little industry. Most of its people live along the Mekong River or near the Tonle Sap (Great Lake) and Tonle Sap River, where conditions are ideal for growing rice. The people live in small villages and work in rice paddies nearby. Their houses are thatched and built on stilts to keep them dry during floods. Rice and fish are the chief foods throughout Cambodia.

Most Cambodians are Khmer, one of the old ethnic groups of Southeast Asia. They speak the Khmer language, which has its own alphabet. Most Cambodians are Buddhists.

About a thousand years ago, the Khmer people controlled a huge Southeast Asian empire. The city of Angkor was the capital of that empire. Angkor now stands in ruins, but its magnificent architecture and sculpture are treasures of Khmer civilization.

Cameroon

Cameroon (25) is located in west-central Africa. It is bordered by Chad, the Central African Republic, Congo, Gabon, Equatorial Guinea, and Nigeria.

Capital: Yaoundé
Official Name: République Unie du Cameroun (United Republic of Cameroon)
Official Languages: French and English
Government: Republic—one-house legislature called the National Assembly; elected president serves as head of government assisted by a cabinet
Flag: Green, red, and yellow vertical stripes, with a yellow star in the center of the red stripe; green stands for the forests of the south, yellow for the savannas of the north, and red for the unity between the two regions
Area: 183,569 sq. mi. (475,442 km²); greatest distances—north-south, 770 mi. (1,239 km); east-west, 450 mi. (724 km); coastline—240 mi. (386 km)
Population: 1983 estimate—9,103,000; distribution—72 per cent rural, 28 per cent urban; density—49 persons per sq. mi. (19 persons per km²)
Largest Cities: (1976 census) Douala (458,246) and Yaoundé (313,706)
Economy: Mainly agricultural
Chief Products: Agriculture—bananas, cassava, cacao, coffee, corn, cotton, livestock, palm oil and kernels, peanuts, sorghum, tea, tobacco; forest products—lumber, rubber; manufacturing—aluminum; mining—petroleum
Money: Basic unit—franc
Climates: Steppe, tropical wet and dry, tropical wet

Cameroon is a black African nation with a population made up of more than 100 different ethnic groups. The groups vary in their culture and traditions, but most rely on farming as a way of life. The people live in simple homes with few modern conveniences, but they grow enough food to feed themselves adequately.

Various groups that live in the forested regions of southern Cameroon grow cassava, yams, and cacao—the chief export crop. Many of the people of the south are Christians. The Fulani people, who live in the hot, dry plains of northern Cameroon, are Muslims. They grow millet and peanuts and tend cattle and horses. The Kirdi people practice traditional African religions. Some are nomads, who move from place to place. Others live in farm settlements where they grow cotton, millet, peanuts, and rice and raise chickens, goats, and sheep. The Bamiléké live in the western mountains, where many of the group grow coffee and practice crafts such as woodworking, weaving, and embroidery.

Canada

Canada (26) is the second largest nation in area in the world. It is located in North America, and it is bordered by the United States.

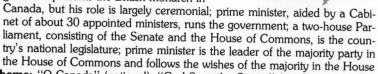

Capital: Ottawa

Official Name: Dominion of Canada

Official Languages: English and French

Government: Constitutional monarchy—governor general represents the British monarch in Canada, but his role is largely ceremonial; prime minister, aided by a Cabinet of about 30 appointed ministers, runs the government; a two-house Parliament, consisting of the Senate and the House of Commons, is the country's national legislature; prime minister is the leader of the majority party in the House of Commons and follows the wishes of the majority in the House

Anthems: "O Canada" (national); "God Save the Queen" (royal)

Flag: A red, 11-pointed maple leaf, the national emblem, appears on a white field; at each end of the flag is a broad, vertical red stripe

Area: 3,831,033 sq. mi. (9,922,330 km²), including 291,571 sq. mi. (755,165 km²) of inland water; greatest distances—east-west, 3,223 mi. (5,187 km), from Cape Spear, Nfld., to Mount St. Elias, Y.T.; north-south, 2,875 mi. (4,627 km), from Cape Columbia on Ellesmere Island to Middle Island in Lake Erie; coastline—151,488 mi. (243,797 km), including mainland and islands; Atlantic Ocean, 28,019 mi. (45,092 km); Arctic Ocean, 82,698 mi. (133,089 km); Hudson Bay, Hudson Strait, and James Bay, 24,786 mi. (39,890 km); Pacific Ocean, 15,985 mi. (25,726 km)

Population: 1983 estimate—24,541,000; distribution—76 per cent urban, 24 per cent rural; density—5 persons per sq. mi. (2 persons per km²)

Largest Cities: (1976 census) Montreal (1,080,546), Toronto (633,318), Winnipeg (560,874), Calgary (469,917), Edmonton (461,361), and Vancouver (410,188)

Economy: Mainly manufacturing, with trade and services contributing significantly

Chief Products: Agriculture—beef cattle, wheat, milk, hay, hogs, rapeseed, poultry, barley, eggs, vegetables, tobacco, fruits; fishing industry—salmon, herring, cod, lobster, scallops; forest industry—logs and bolts, pulpwood; fur industry—mink, beaver, fox, lynx; manufacturing—food products, transportation equipment, paper products, fabricated metal products, primary metals, chemicals, lumber and wood products, electric machinery and equipment, nonelectric machinery, printed materials, stone, clay, and glass products, textiles, clothing; mining—petroleum, natural gas, iron ore, copper, natural gas liquids, zinc, coal, nickel, uranium, asbestos, potash, sand and gravel

Money: Basic unit—dollar

Climates: Extremely varied—polar, subarctic, continental moist, oceanic moist, steppe, highlands

Canada is a huge but thinly populated country. It is a federation, or union, of 10 provinces and two territories. Each province has its own government. While the two territories have some self-government, they are controlled largely by the country's central government. The Atlantic Provinces, on the Atlantic Ocean, include Newfoundland, New Brunswick, Nova Scotia, and Prince Edward Island. They have about 10 per cent of the nation's population. The Prairie Provinces, in the central portion of the country, hold about 17 per cent of Canada's people and include Alberta, Saskatchewan, and Manitoba. Other provinces are Quebec, Ontario, and British Columbia, which together have most of the nation's population. The two territories to the north are the Yukon Territory and the Northwest Territories. Because of the territories' severe climate and remote location, however, less than 1 per cent of Canada's people live there.

Canada is a land of immense variety. The natural beauty of the country's landscape ranges from towering mountains to crystal-clear lakes, and from lush green forests to vast prairies and wheat fields. To the north is the severe contrast of Arctic wastelands. All this provides magnificent scenery for visitors and citizens alike.

The country's wealth of natural resources makes it one of the world's prosperous nations. Fertile farmland, huge mineral deposits, vast forests, and rich fishing waters contribute to Canada's well-developed economy. Foreign markets eager for Canadian minerals have accounted for the settlement of the country's more remote areas. Canada is now the world's leading exporter of minerals.

About two-thirds of Canadian workers are employed in industries providing community, business, and personal services such as schools, hospitals, financial institutions, advertising agencies, and restaurants. Manufacturing is also important and is divided into two broad fields: processing minerals and other natural resources for export, and providing products for use by Canadians.

Indians and Eskimos were the original people of Canada and make up about 2 per cent of the present population. Most Canadians, however, are of European descent. About 45 per cent have British ancestry, and nearly 30 per cent are descendants of early French settlers. Canada's population has almost doubled since the end of World War II in 1945 due largely to heavy immigration. Canada has become a new home for political refugees from many nations. In 1956, for example, thousands of Hungarians came to Canada after the revolution in their country. From 1975 to 1980, roughly 60,000 refugees from Cambodia, Laos, and Vietnam settled in Canada.

French and English are the country's official languages. About 67 per cent of Canadians speak only English. About 18 per cent speak only

French, and approximately 13 per cent speak both languages. Most of the French-speaking Canadians live in the province of Quebec. As a result, Quebec differs somewhat from other areas of Canada. Rural areas of Quebec, in particular, reflect the strong influence of French culture. The homes and customs of the people indicate old French traditions. Montreal is Quebec's largest city and the center of its economic and cultural life.

The rapid development of Canada's manufacturing industries since the 1940's has caused a population shift from rural to urban areas. But the decreasing supply of Canada's energy resources has caused many citizens to oppose such rapid urban growth. More than three-fourths of Canada's population now live in cities and towns. Most engage in a wide variety of cultural and recreational activities including plays, concerts, numerous athletic events, and visits to museums and parks. Expressway systems linking cities and suburbs, skyscrapers, and apartment buildings are all a part of city scenery.

About 25 per cent of Canadians live in rural communities, but only 6 per cent are farmers. Others work at fishing, mining, lumbering, and some commute to jobs in the cities. Most farms lie in the Prairie Provinces. Farming is generally a family activity, and modern equipment has facilitated the jobs of farming families. A typical farm consists of a wooden farmhouse surrounded by buildings where grain and machinery are stored. Community life centers around the small market towns that have grown up along the railroads. The railroads are a necessary link to ports on the Great Lakes and the Pacific Ocean. Farm people shop, participate in community activities, and attend school and church in these nearby towns.

Canada's broad Arctic region has a sparse population, a third of which is made up of Indians and Eskimos. They have inhabited the area for thousands of years and many still follow their traditional occupations of fishing, hunting, and trapping. But the Arctic's old ways of life have largely ended. Tents and igloos, dog sleds and kayaks are no longer the rule. Most people now live in modern houses, wear modern clothing, eat food bought in stores, and travel by snowmobile or motorboat. This end of traditional life style has brought about many social problems—high rates of unemployment, alcoholism, and crime. The discovery of petroleum in the Arctic region, though, has created hope for the future.

The majority of Canadians are Roman Catholics. Early French settlers brought the Roman Catholic faith to Canada, and the church has remained important in education and social life. Most other Canadians are Protestants. In addition, there are small numbers of Jews, Muslims, Buddhists, Hindus, and Sikhs. Canada's earliest schools were operated by religious groups. But in 1867, education became the responsibility of the provincial governments. In some provinces, the law provides for the public school system to include separate schools for certain religious groups.

Canada's history is basically a story of years of struggle between the French and the British. Indian tribes were the area's first inhabitants before the arrival of European explorers, missionaries, and fur trappers during the 1500's and 1600's. The French were the first colonists to settle in the area. In 1608, Samuel de Champlain founded Quebec, the first permanent settlement in Canada. In 1663, Canada became a province of France and was called New France. Around 1689, the French began fighting with British colonists from New England over the area's rich fur trade. Struggles and wars between the two groups continued until the British were victorious. The Treaty of Paris was signed in 1763, turning Canada over to Great Britain. More than 100 years passed before the four provinces of Quebec, Ontario, New Brunswick, and Nova Scotia became the Dominion of Canada under the British North America Act in 1867. Other provinces and territories gradually joined the Dominion, the last being Newfoundland in 1949. The name Canada probably came from an Iroquois Indian word meaning village, community, or group of huts.

Canadians enjoy sports of all sorts. Lacrosse was the country's first national game. The Indians played it before the arrival of the Europeans. But ice hockey is now the national pastime. Canadians can begin competing in amateur hockey leagues at the age of 7. Other popular spectator sports include football, baseball, and soccer. Favorite recreational activities during the long Canadian winters include skiing, snowshoeing, skating, and tobogganing. Numerous ski slopes throughout the mountains of the Canadian Rockies attract thousands of skiers every year. Popular summertime activities are swimming, hiking, fishing, tennis, and golf.

Fine arts are an important part of Canadian life. The Stratford Festival, held annually in Stratford, Ontario, is one of the popular Canadian theatrical events. Theater performances flourish in both English- and French-speaking areas of the country. In addition, ballet, opera, music, painting and sculpture, motion pictures, and architecture have all vastly developed with the support of the country's government. Among the fine works of Canadian sculpture are the ivory and soapstone carvings by Eskimo artists. They depict objects and activities from the daily life of Eskimos. Examples of many types of architecture can be seen throughout Canada. Traditional styles are exhibited by the French-style homes of Quebec and the neo-Gothic Parliament buildings in the capital city of Ottawa. More modern designs include the Dominion Centre and the City Hall in Toronto, and the Place Ville Marie in Montreal.

Cape Verde

Cape Verde (27) is an island country composed of 10 islands and 5 is-
lets in the Atlantic Ocean. It lies
north of the equator, off the west
coast of Africa.

Capital: Praia
Language: Most speak a local Creole
    dialect based on ancient Portuguese
    and various African languages
Government: Republic—elected legislature, the
    People's Assembly, selects a president and a premier and appoints an eight-
    member Cabinet headed by the premier
Flag: A black star and a yellow seashell, framed by two curved cornstalks, lie on
    a red vertical stripe at the left; a yellow horizontal stripe appears over a green
    one at the right
Area: 1,557 sq. mi. (4,033 km^2); coastline—600 mi. (966 km)
Population: 1983 estimate—342,000; distribution—80 per cent rural, 20 per
    cent urban; density—220 persons per sq. mi. (85 persons per km^2)
Largest Cities: (1970 census) Mindelo (28,797) and Praia (21,494)
Economy: Based mainly upon agriculture and fishing, with limited mining
Chief Products: Bananas, salt, sugar cane
Money: Basic unit—escudo
Climate: Steppe

Cape Verde has few natural resources. Its mountainous islands are covered largely by volcanic ash. The infertile soil and extremely dry climate
make growing crops difficult, and, as a result, Cape Verdeans have a low
standard of living. Many are unemployed year-round. When the rainfall
is adequate, others make a bare living by fishing or by growing crops
such as coffee, sugar cane, fruits, and vegetables.

Portugal ruled Cape Verde for more than 500 years before the island
nation became independent in 1975. About 70 per cent of the people
are of mixed Portuguese and black African descent. The majority of the
rest are black Africans. Most Cape Verdeans speak a Creole dialect that
combines African languages and ancient Portuguese. About three-fourths
of the people can read and write.

Cape Verde has about 920 miles (1,480 kilometers) of roads, but no
railroads. It has one airport and several small landing strips. Boats travel
among the islands infrequently.

Central African Republic

The Central African Republic (28)
is located in the center of Africa.
It is bordered by Chad, Sudan, Zaire,
Congo, and Cameroon.

Capital: Bangui
Official Name: Le République Cen-
trafricaine (The Central African Republic)
Official Language: French
Government: Republic—army officers overthrew
civilian government in 1981; a committee made up of army officers now
heads government
Flag: Horizontal blue, white, green, and yellow stripes are divided at the center
by a red vertical stripe; yellow star represents the guiding light of the future;
red, white, and blue recall the French flag; green, yellow, and red are for the
people and their unity
Area: 240,535 sq. mi. (622,984 km^2); greatest distances—east-west, 890 mi.
(1,424 km); north-south, 590 mi. (944 km)
Population: 1983 estimate—2,385,000; distribution—59 per cent rural, 41 per
cent urban; density—10 persons per sq. mi. (4 persons per km^2)
Largest Cities: (1974 est.) Bangui (301,000) and Berbérati (93,000)
Economy: Mainly farming, with some mining and very limited manufacturing
Chief Products: Agriculture—bananas, coffee, cotton, livestock, palm kernels,
peanuts, rubber, sesame, yams; forestry—timber; mining—diamonds, gold
Money: Basic unit—franc
Climates: Steppe, tropical wet and dry, tropical wet

The Central African Republic is a poor, landlocked nation with few natural resources. Tourists come to hunt or to photograph antelope, elephants, gorillas, lions, rhinoceroses, and other animals that run wild over
the grassy plateaus that cover much of the country.

Most of the people of the Central African Republic are black Africans.
About three-fifths of them live in rural areas. Some work on small plantations where coffee, cotton, and rubber are grown for export; but most
of the people simply hunt, fish, grow food crops, and gather insects and
caterpillars as food. They also raise goats, pigs, poultry, and sheep. Cattle
cannot be raised in many parts of the country because of the presence
of tsetse flies, which cause an illness called sleeping sickness. Sleeping
sickness kills cattle and often kills people as well.

The Central African Republic has no railroads, and many of its roads
are impassable during the rainy season. Rivers provide the chief transportation routes. Diamonds are the nation's only major mineral resource.

Chad

Chad (29) is located in north-central Africa. It is bordered by Libya, Sudan, the Central African Republic, Niger, Nigeria, and Cameroon.

Capital: N'Djamena
Official Name: République du Tchad (Republic of Chad)
Official Language: French
Government: Republic—president heads government assisted by a 20-member council
Flag: Vertical blue, yellow, and red stripes; blue symbolizes the sky and hope; yellow stands for the sun; red represents fire and unity
Area: 495,755 sq. mi. (1,284,000 km^2); greatest distances—east-west, 680 mi. (1,088 km); north-south, 1,070 mi. (1,712 km)
Population: 1983 estimate—4,843,000; distribution—82 per cent rural, 18 per cent urban; density—10 persons per sq. mi. (4 persons per km^2)
Largest Cities: (1975 est.) N'Djamena (224,000) and Sarh (50,000)
Economy: Based almost entirely on agriculture
Chief Products: Agriculture—beans, cassava, cotton, dates, livestock (cattle, goats, sheep), millet, peanuts, rice, sorghum, tropical fruit, wheat; mining—natron (washing soda), tungsten; fishing—fresh and dried fish
Money: Basic unit—franc
Climates: Steppe, desert, tropical wet and dry

Chad is an underdeveloped nation covered by desert in the north and grassy plains in the south and central areas. Sharp differences also exist between the inhabitants of the north and the south. Chad's population consists of about 50 different ethnic groups. The people of the north are Muslims. Most are Arabs or persons of Hamitic origin. Many are nomads who live in tents and roam across the desert in search of water and grazing land for their livestock.

In the south, many of the people live in mud huts with thatched roofs. Most are black Africans who farm for a living. Cotton is Chad's chief export, and the people of the south also grow food crops such as sorghum, millet, rice, beans, and cassava. Some southerners are Christians, but most practice traditional African religions.

French is the official language of Chad, but each ethnic group speaks its own language or dialect. About a third of the children of Chad receive some elementary education. Only about 18 per cent of the people can read and write.

Chile

Chile (30) is located on the southwest coast of south America. It is
bordered by Peru, Argentina, and Bolivia.

Capital: Santiago
Official Name: República de Chile
(Republic of Chile)
Official Language: Spanish
Government: Military rule—four-member military
junta or council rules the country; leader of
junta serves as president and holds policymaking authority
National Anthem: "Cancion Nacional de Chile" ("National Anthem of Chile")
Flag: Bottom half red; top half has a blue square to the left with the remainder
white; single white star in center of blue stands for unity; red is for the blood
of heroes, white for the snow of the Andes, and blue for the sky
Area: 292,258 sq. mi. (756,945 km^2); greatest distances—north-south, 2,650 mi.
(4,265 km); east-west, 225 mi. (362 km); coastline—3,317 mi. (5,338 km)
Population: 1983 estimate—11,675,000; distribution—80 per cent urban, 20
per cent rural; density—39 persons per sq. mi. (15 persons per km^2)
Largest Cities: (1980 est.) Santiago (3,899,495), Viña del Mar (277,068), Val-
paraíso (266,354), and Concepción (200,602)
Economy: Mainly mining and manufacturing
Chief Products: Agriculture—barley, beans, cattle, corn, flax, fruits, hemp, lentils, milk, oats, peas, pigs, potatoes, rice, rye, sheep, sunflowers, tobacco,
wheat, wines; manufacturing and processing—bakery goods, beef, blankets,
canned food, cement, clothing, drugs, flour, glassware and pottery, household appliances, leather goods, paper and paper containers, rubber goods,
shoes, soaps, stoves, textiles, tires; mining—coal, copper, gold, iodine, iron
ore, lead, nitrates, petroleum, silver, zinc; forest products—pinewood, pulp
Money: Basic unit—peso; 100 centésimos equal 1 peso
Climates: Varied—oceanic moist, subtropical dry summer, desert, highlands

Chile is known for its rich mineral deposits—especially copper—and for
the beauty of the Andes Mountains, which run down the length of the
country. About 90 per cent of all Chileans live in a fertile valley between
the Andes in the east and coastal mountains in the west.

Nearly all Chileans have mixed Spanish and Indian ancestry or are
descendants of Spanish or other European settlers. Most Chilean workers
are farmers, factory laborers, or miners. Cowboys called *huasos* work on
the country's large cattle ranches.

Many Chileans eat and dress well and have comfortable homes, but
many others have a very low standard of living. Nearly 90 per cent of
the people can read and write. However, most Chileans do not attend
school beyond the elementary grades.

China

China (31) is located in eastern and central Asia. It is bordered by Russia, Afghanistan, India, Nepal, Pakistan, Bhutan, and a number of other Asian nations.

Capital: Peking
Official Name: Chung-hua Jen-min Kung-
ho-kuo (People's Republic of China)
Official Language: Chinese (Northern dialect)
Government: Communist dictatorship—dominated by three organizations: Chinese Communist Party, the military, and the State Council, a branch of the government; the Communist Party strictly controls the political system; the State Council carries on day-to-day affairs of government and is led by premier
National Anthem: "The East is Red"
Flag: Red with one large yellow star and four small stars in top left corner
Area: 3,678,470 sq. mi. (9,527,200 km^2); greatest distances—north-south, 2,500 mi. (4,023 km); east-west, 3,000 mi. (4,828 km); coastline—4,019 mi. (6,468 km), including 458 mi. (737 km) for Hai-nan Island
Population: 1983 estimate—1,025,874,000 (based on UN data); distribution—80 per cent rural, 20 per cent urban; density—280 persons per sq. mi. (108 per km^2)
Largest Cities: (1970 est.) Shanghai (10,820,000), and Peking (7,570,000)
Economy: Mainly agricultural, with developing industry
Chief Products: Agriculture—rice, wheat, cotton, corn, tea, tobacco, sorghum, hogs, barley, millet, peanuts; manufacturing—iron and steel, machinery, textiles; mining—coal, iron ore, petroleum, tungsten, antimony
Money: Basic unit—yuan
Climates: Extremely varied—subarctic, continental moist, subtropical moist, steppe, desert, highlands, tropical wet and dry, tropical wet

China has more people than any other nation. It has at least 14 cities with more than a million residents, but about 80 per cent of its people live in rural areas.

The Chinese government owns the factories, controls communication and education, rations certain foods and clothing materials, and influences many other aspects of life in China. Almost all adults work outside the home. In many families, men and women share household chores and child-rearing responsibilities. Most Chinese live in simple two- or three-room houses or small apartments.

The Chinese are proud of their ancient civilization's contributions to the arts and sciences. Education is highly valued, and more than 90 per cent of all adults can read and write.

Colombia

Colombia (32) is located in the northwestern corner of South America. It is bordered by Venezuela, Brazil, Peru, Ecuador, and Panama.

Capital: Bogotá
Official Name: República de Colombia (Republic of Colombia)
Official Language: Spanish
Government: Republic—elected president appoints 13 members to Cabinet; two-house Congress elected by people
National Anthem: "El Himno Nacional"
Flag: A yellow horizontal stripe above narrower blue and red stripes
Area: 439,737 sq. mi. (1,138,914 km^2); greatest distances—northwest-southeast, 1,170 mi. (1,883 km); northeast-southwest, 850 mi. (1,368 km); coastline—580 mi. (933 km) along the Pacific Ocean; 710 mi. (1,143 km) along the Caribbean Sea.
Population: 1983 estimate—29,810,000; distribution—60 per cent urban, 40 per cent rural; density—67 persons per sq. mi. (26 persons per km^2)
Largest Cities: (1973 census) Bogotá (2,850,000) and Medellín (1,064,741)
Economy: Mainly agricultural, with some mining and manufacturing
Chief Products: Agriculture—bananas, cacao, coffee, corn, manioc, potatoes, rice, sugar cane, wheat; mining—coal, emeralds, gold, petroleum, platinum, salt, silver; manufacturing—beverages, cement, chemicals, leathers, steel, sugar, textiles, tobacco
Money: Basic unit—peso
Climates: Steppe, highlands, tropical wet and dry, tropical wet

Colombia is a land of rugged mountains, fertile valleys, dense forests, and tropical jungles. Agriculture is the most important economic activity, though only about 4 per cent of the country's land area is cultivated. Farms and small plantations produce coffee, bananas, cacao, beans, rice, sugar cane, wheat, and other crops.

Colombia's natural resources include large oil fields and precious minerals such as gold, silver, platinum, and emeralds. The country's developing manufacturing industries include factories that make food products, chemicals, shoes, and textiles.

Most Colombians are either of Spanish or other European ancestry, or of mixed European and Indian ancestry. Indians account for about 7 per cent of the population. Most of the people are Roman Catholics. About 60 per cent of all Colombians live in cities and towns. Some of the cities rank among the older settlements in the Western Hemisphere. Historic churches, fortresses, and Spanish-style homes stand among modern structures.

Comoros

Comoros (33) is located off the
east coast of Africa. It is composed of several islands located
in the Indian Ocean.

Capital: Moroni
Official Name: Federal and Islamic
Republic of the Comoros
Official Language: French
Government: Republic—elected presi-
dent appoints prime minister and Cabinet; members
of legislative body, called Federal Assembly, elected by people
Flag: A green field covers the flag; crescent moon and four five-pointed stars are
in the center; the green color and the crescent symbolize Islam; the four stars
represent the four islands of the country
Total Land Area: 838 square miles (2,171 km²); coastline—243 mi. (391 km)
Population: 1983 estimate—358,000; distribution—88 per cent rural, 12 per
cent urban; density—427 persons per sq. mi. (165 persons per km²)
Largest City: (1976 est.) Moroni (19,778)
Economy: Almost entirely agricultural
Chief Products: Bananas, cloves, coconuts, corn, perfume plants, rice, spices,
vanilla, yams
Money: Basic unit—franc
Climate: Tropical wet and dry

Comoros is a poor, undeveloped island nation with few natural resources
or industries. Almost all the people are farmers, but they do not grow
enough to feed themselves. They eat bananas, coconuts, and yams and
import large amounts of rice, their chief food. Rain provides the only
natural source of drinking water. Comoros exports cloves and other
spices, perfume plants, and vanilla.

Most Comorians have a mixed ancestry. They are descended from
Arabs, black Africans, and other groups. Most are Muslims and speak
Arabic or Swahili. Many Comorians understand French, the country's
official language, but few speak or write it.

Comoros has a shortage of doctors and hospitals. Illness, disease, and
malnutrition are serious problems along with hunger and lack of education. Only about a third of the country's children finish elementary
school. Comoros has one airport and one radio station. The islands have
a total of about 150 miles (241 kilometers) of paved roads.

Congo

Congo (34) is located in west-central Africa. It is bordered by Cameroon, the Central African Republic, Zaire, Angola, and Gabon.

Capital: Brazzaville
Official Name: République Populaire du Congo (People's Republic of the Congo)
Official Language: French
Government: Republic—controlled by 11-member military council which appoints a president to serve as head of state
Flag: A golden-yellow five-pointed star lies above crossed hammer and hoe in the upper left corner of a red field; green palm branches surround the star, hammer, and hoe
Area: 132,047 sq. mi. (342,000 km^2); greatest distances—north-south, 590 mi. (950 km); east-west, 515 mi. (829 km); coastline—100 mi. (160 km)
Population: 1983 estimate—1,660,000; distribution—63 per cent rural, 37 per cent urban; density—13 persons per sq. mi. (5 persons per km^2)
Largest Cities: (1972 est.) Brazzaville (250,000); (1970 prelim. census) Pointe-Noire (135,000) and Dolisie (25,000)
Economy: Based mainly on agriculture and forestry, with some mining
Chief Products: Agriculture—bananas, cassava, coffee, palm kernels and oil, peanuts, plantains, rice, rubber, sugar cane, sweet potatoes, yams; forestry—limba, mahogany, okoumé; mining—potash
Money: Basic unit—franc
Climates: Tropical wet and dry, tropical wet

Congo is a hot, humid country with few natural resources or industries. It serves, however, as an important transportation center. Pointe-Noire is an Atlantic port that handles overseas trade for Gabon and for the inland countries of Chad and the Central African Republic. Passengers and freight travel on the Congo and Ubangi rivers and on the Congo-Ocean railroad, which links Brazzaville with Pointe-Noire.

Jungles cover more than half the Congo's territory. Only wild animals live in much of this region. Most of the people live on the country's southern border or on the coast. Nearly all Congolese are black Africans. Most of them are farmers who grow bananas, corn, rice, and other food crops. Many others hunt and fish for a living. Lumber is the Congo's chief export.

About half of the Congo's people practice traditional African religions. A small number are Muslims. Most of the rest are Christians. Most adults in the Congo cannot read or write, but about three-fourths of the children receive some elementary education.

Costa Rica

Costa Rica (35) is located in Central America. It is bordered by Nicaragua and Panama.

Capital: San José
Official Name: República de Costa Rica (Republic of Costa Rica)
Official Language: Spanish
Government: Republic—elected president and members of Cabinet make up Council of Government which conducts foreign affairs and enforces federal laws; one-house legislature is called Legislative Assembly
National Anthem: "Himno Nacional" ("National Anthem")
Flag: Two horizontal blue stripes across top and bottom; a wide horizontal red stripe through middle of flag; two white stripes separate red from blue; coat of arms lies on red stripe
Area: 19,575 sq. mi. (50,700 km²); greatest distances—north-south, 220 mi. (354 km); east-west, 237 mi. (381 km); coastline—380 mi. (612 km) on the Pacific Ocean; 133 mi. (214 km) on the Caribbean Sea
Population: 1983 estimate—2,411,000; distribution—59 per cent rural, 41 per cent urban; density—124 persons per sq. mi. (48 persons per km²)
Largest Cities: (1973 census) San José (215,441) and Alajuela (30,190)
Economy: Mainly agricultural, with rapidly growing industry
Chief Products: Agriculture—bananas, cacao, cattle, coffee, corn, sugar cane; manufacturing—furniture, leather goods, processed foods, textiles
Money: Basic unit—colón
Climates: Tropical wet and dry, tropical wet

Costa Rica is a small, mountainous country covered largely with fertile, volcanic soil. About three-fifths of all Costa Ricans live in rural areas, and agriculture employs about half the country's workers. Coffee and bananas rank as the chief exports. Farmers also raise livestock and grow sugar, cacao, corn, and other crops.

Most Costa Rican farmers live in brightly painted wooden houses or in small adobe houses with tile roofs. Most city dwellers live in apartments or row houses. Everyday meals in Costa Rica often include beans, corn, coffee, eggs, rice, and various tropical fruits. On festive occasions, boiled iguana eggs and cooked hog's head may be served as special treats.

Nearly all Costa Ricans are whites or are of mixed Spanish and Indian ancestry. Most of them speak Spanish, and about 90 per cent are Roman Catholics. Some religious holidays are celebrated with colorful festivals, which may include bullfights, fireworks, and parades. Soccer is the country's national sport.

Cuba

Cuba (36) is located in the Caribbean Sea. It lies in the West Indies, about 90 miles south of Florida.

Capital: Havana
Official Language: Spanish
Government: Socialist republic (Communist dictatorship)—most powerful official is president of the Council of State; government strongly controlled by prime minister and the Communist Party of Cuba
National Anthem: "La Bayamesa"
Flag: Alternating blue and white horizontal stripes; a white star symbolizing independence lies on a red triangle near the staff
Area: 44,218 sq. mi. (114,524 km^2); greatest distances—northwest-southeast, 759 mi. (1,221 km); north-south, 135 mi. (217 km); coastline—2,100 mi. (3,380 km)
Population: 1983 estimate—10,252,000; distribution—60 per cent urban, 40 per cent rural; density—231 persons per sq. mi. (89 persons per km^2)
Largest Cities: (1975 est.) Havana (1,900,240); (1970 census) Santiago de Cuba (275,970) and Camagüey (196,854)
Economy: Based mainly on services, agriculture, and industry
Chief Products: Agriculture—cattle, citrus fruits, coffee, pineapples, sugar cane, tobacco, vegetables; manufacturing—cement, cigarettes, cigars, fertilizers, refined sugar, rum, textiles; mining—chromite, iron, limestone, manganese, nickel
Money: Basic unit—peso
Climates: Tropical wet and dry, tropical wet

Cuba is a beautiful island nation with a mild climate and fertile soil. Sugar grows throughout most of Cuba and is the country's chief crop. Tobacco, another important crop, is made into world-famous Cuban cigars.

Most Cubans are either whites of Spanish descent or are of mixed black and white ancestry. Almost all Cubans speak Spanish, and some also speak English. About 60 per cent of the people live in urban areas. Many Cuban farmers work on large, government-operated state farms. Other farmers own their land but must sell their crops to the government.

Many Cuban city dwellers live in small apartments, sometimes shared by two or more families. Thatch-roofed huts provide housing for rural people. Most Cubans have enough to eat, though such foods as eggs and meat are scarce. The government provides free education and medical care. For recreation, Cubans enjoy singing and dancing, baseball, and other sports.

Cyprus

Cyprus (37) is an island nation located in Southwest Asia at the eastern end of the Mediterranean Sea.

Capital: Nicosia
Official Name: In Greek, Kypriaki Dimokratia; in Turkish, Kibris Cumhuriyeiti (Republic of Cyprus)
Official Languages: Greek and Turkish
Government: Republic—president serves as head of state and government; Turkey and Turkish Cypriots, though, refuse to recognize the Greek Cypriot government
National Anthem: "Imnos pros tin Eleftherian" ("The Hymn to Liberty")
Flag: White with a map of Cyprus in copper-yellow (for copper) in the center above two green crossed olive branches (for peace)
Area: 3,572 sq. mi. (9,251 km^2); greatest distances—east-west, 128 mi. (206 km); north-south, 75 mi. (121 km)
Population: 1983 estimate—635,000; distribution—58 per cent rural, 42 per cent urban; density—179 persons per sq. mi. (69 persons per km^2)
Largest Cities: (1974 est.) Limassol (55,000) and Nicosia (51,000)
Economy: Based mainly on tourism, agriculture, manufacturing, and mining
Chief Products: Agriculture—barley, carrots, grapefruit, grapes, lemons, oranges, potatoes, wheat; manufacturing—cigarettes, olive oil, plastics, shoes, textiles, wines; mining—asbestos, copper, iron
Money: Basic unit—Cyprus pound
Climate: Subtropical dry summer

Cyprus is a scenic nation noted for its sandy beaches, rocky mountains, hilltop castles, and old churches. About 80 per cent of its people are of Greek origin, and most of the rest are of Turkish origin. Each group speaks its own language and has its own schools. Most of the Greeks are Christians, and most of the Turks are Muslims.

Tourism is the most important industry in Cyprus, although mining is also important. Asbestos, copper, and iron ore are its chief minerals. Fertile soil, a warm climate, and irrigation systems also make Cyprus a productive agricultural nation.

Slightly more than half of all Cypriots live in rural areas. Most people live in simple homes built around a courtyard. Modern apartment buildings provide housing for many city dwellers. Older sections of the cities have small, open-air shops linking narrow streets. Cyprus has no railroads, but good highways link most parts of the island.

Czechoslovakia

Czechoslovakia (38) is located in central Europe. It is bordered by Poland, Russia, Hungary, Austria, and East and West Germany.

Capital: Prague

Official Name: Československá Socialistická Republika (Czechoslovak Socialist Republic)

Official Languages: Czech and Slovak

Government: Socialist republic (Communist dictatorship)—two-house legislature, Federal Assembly, elects president who appoints prime minister and Cabinet; however, members of Communist Party hold all important government positions

National Anthem: "Kde domov můj?" ("Where Is My Homeland?"), combined with "Nad Tatrou sa blýská" ("Lightning Flashes over the Tatra")

Flag: Divided into three sections; top half white, bottom half red, and a blue triangle to the left near staff

Area: 49,370 sq. mi. (127,869 km^2); greatest distances—east-west, 470 mi. (756 km); north-south, 235 mi. (378 km)

Population: 1983 estimate—15,682,000; distribution—67 per cent urban, 33 per cent rural; density—319 persons per sq. mi. (123 persons per km^2)

Largest Cities: (1978 est.) Prague (1,188,573) and Brno (369,028)

Economy: Mainly manufacturing, with some agriculture

Chief Products: Agriculture—barley, cattle, hogs, hops, potatoes, sugar beets, wheat; manufacturing—chemicals, glass products, machinery, military equipment, textiles, transportation equipment; mining—coal

Money: Basic unit—koruna

Climate: Continental moist

Czechoslovakia is one of Eastern Europe's highly industrialized nations. Many of its factories produce heavy machinery and transportation equipment. Czechoslovakia is also well known for two other products: beer and delicate crystal.

Farmland covers much of Czechoslovakia. Most farmers work on government-controlled state farms or collective farms.

Two closely related Slavic groups, the Czechs and the Slovaks, make up most of the country's population. The people have one of the high standards of living in Eastern Europe. Nearly all the people can read and write. Many families own a television and refrigerator, and about a fifth own a car. Most city dwellers live in apartments, and most rural families live in small, modern houses. Czechoslovaks enjoy music and have a rich tradition of folk songs and dances. Popular recreational activities include soccer games and motion pictures.

Denmark

Denmark (39) is located in northern Europe. It is bordered by West Germany.

Capital: Copenhagen
Official Name: Kongeriget Danmark (Kingdom of Denmark)
Official Language: Danish
Government: Constitutional monarchy—king or queen acts as head of state; prime minister serves as head of government and forms a cabinet called the Council of State; one-house parliament is called the *Folketing*
National Anthems: "Kong Christian stod ved højen mast" ("King Christian Stood by Lofty Mast") and "Der er et yndigt land" ("There Is a Lovely Land")
Flag: Red with two white stripes—one horizontal and one vertical
Area: 16,629 sq. mi. (43,069 km²); greatest distances—east-west, 250 mi. (402 km); north-south, 225 mi. (362 km); coastline—1,057 mi. (1,701 km)
Population: 1983 estimate—5,169,000; distribution—84 per cent urban, 16 per cent rural; density—311 persons per sq. mi. (120 persons per km²)
Largest Cities: (1978 est.) Copenhagen (515,594) and Århus (245,386)
Economy: Mainly manufacturing
Chief Products: Agriculture—bacon, barley, beef and dairy cattle, beets, eggs, hogs, oats, potatoes, poultry, rye, wheat; fishing—cod, haddock, herring, plaice, salmon, trout; manufacturing—cement, diesel engines, electrical equipment, furniture, machinery, processed foods, ships, silverware
Money: Basic unit—krone
Climates: Continental moist, oceanic moist

Denmark is a prosperous nation, though it has few natural resources. To pay for the raw materials it must import for its industries, Denmark exports finely crafted manufactured products such as furniture, porcelain, and silverware. Its small, well-kept farms produce high-quality butter, cheeses, bacon, and hams. Danes have been a seafaring people since the days of the Vikings, and shipping and fishing industries are still important to Denmark's economy.

Danes have a high standard of living. Farmland covers about three-fourths of Denmark, but about 85 per cent of its people live in cities and towns. Danish cities have modern housing and offices as well as charming, well-preserved older sections of picturesque buildings and cobblestone streets. Danes eat a variety of foods, and they often serve elaborate, open-faced sandwiches called *smørrebrød* for lunch or a late-evening supper.

Djibouti

Djibouti (40) is located in eastern Africa. It lies on the western shore of the Gulf of Aden, and it is bordered by Ethiopia and Somalia.

Capital: Djibouti
Formerly Called: French Somaliland; French Territory of the Afars and Issas
Official Language: Arabic
Government: Republic—one-house legislature, called National Assembly, elected by people; assembly elects a president who heads the government
Flag: A blue horizontal stripe at the top, a green horizontal stripe at the bottom, and a red star on a white triangle near the staff
Area: 8,494 sq. mi. (22,000 km^2); greatest distances—east-west, 110 mi. (177 km); north-south, 125 mi. (201 km); coastline—152 mi. (245 km)
Population: 1983 estimate—355,000; distribution—53 per cent urban, 47 per cent rural; density—41 persons per sq. mi. (16 persons per km^2)
Largest City: (1973 est.) Djibouti (100,000)
Economy: Based almost entirely on shipping and railway transportation
Chief Products: Hides, skins
Money: Basic unit—Djibouti franc
Climate: Desert

Djibouti is a small, desolate country with no valuable natural resources and little industry. Its economy depends almost entirely on the port in the capital city of Djibouti and on a railroad that links the port with Addis Ababa, Ethiopia. Djibouti handles much of Ethiopia's trade.

Two ethnic groups, the Afars and the Issas, make up most of Djibouti's population. A large number maintain their traditional way of life as nomads, wandering over the countryside with their herds. The country's intense heat, scarcity of water, and shortage of grazing land, however, make it almost impossible for the people to earn a living. A 90 per cent unemployment rate plagues the city of Djibouti, which is home for about one-third of the country's population.

Nearly all the people of Djibouti are Muslims. Arabic is the country's official language, but most of the people speak Afar or Somali. Only about 10 per cent of the people can read and write. Djibouti has one international airport and about 100 miles (160 kilometers) of paved roads.

Dominica

Dominica (41) is an island nation in the Caribbean Sea. It lies 320 miles (515 kilometers) north of the Venezuelan coast.

Capital: Roseau
Official Name: Commonwealth of Dominica
Official Language: English
Government: Republic—one-house legislature, called the House of Assembly, makes nation's laws and elects prime minister from among its members; prime minister sits on an eight-member Cabinet, which conducts operations of government
Flag: Dark green with three horizontal stripes and three vertical stripes of yellow, white, and black forming a cross; a large red circle in center of cross rimmed with ten small yellow stars and centered with a blue eagle
Area: 290 sq. mi. (751 km^2); greatest distances—east-west, 16 mi. (25.6 km); north-south, 30 mi. (48 km); coastline—89 mi. (148 km)
Population: 1983 estimate—80,000; distribution—80 per cent rural, 20 per cent urban; density—277 persons per sq. mi. (107 persons per km^2)
Largest City: (1976 est.) Roseau (10,157)
Economy: Based on the export of agricultural goods, with some food processing, tourism, and mining
Chief Products: Agriculture—bananas, coconuts
Money: Basic unit—East Caribbean dollar
Climate: Tropical wet

Dominica is a tiny island nation that depends largely on agriculture to support its people. More than 60 per cent of them work on farms, and most of the rest work in industries that process agricultural products. Bananas rank as the leading crop and export. Coconuts and coconut by-products are also important.

Most Dominicans have African or mixed African, British, and French ancestry. A small number are descended from Carib Indians. The majority of Dominican city dwellers speak English, the country's official language. In rural areas, where about four-fifths of the population lives, most people speak a language called French patois, which combines African languages and French. About 80 per cent of the people are Roman Catholics, and nearly all the others are Protestants.

The Caribbean Sea provides Dominicans with crayfish, crabs, and lobsters. Other common foods in Dominica include frog legs, sweet potatoes, and bananas.

Dominican Republic

The Dominican Republic (42) makes up the eastern two-thirds of the island of Hispaniola between the Atlantic Ocean and the Caribbean Sea.

Capital: Santo Domingo
Official Language: Spanish
Government: Republic—headed by elected president who appoints a Cabinet; two-house legislature is also elected by the people
National Anthem: "Himno Nacional"
Flag: A white cross divides the flag into quarters which are alternately red and blue; the Dominican coat of arms is centered on the cross; blue stands for liberty, white for salvation, and red for the blood of heroes
Area: 18,816 sq. mi. (48,734 km²); greatest distances—east-west, 240 mi. (388 km); north-south, 170 mi. (274 km); coastline—604 mi. (972 km)
Population: 1983 estimate—5,935,000; distribution—51 per cent rural, 49 per cent urban; density—316 persons per sq. mi. (122 persons per km²)
Largest Cities: (1976 est.) Santo Domingo (979,608) and Santiago (219,846)
Economy: Mainly agricultural, with some manufacturing
Chief Products: Agriculture—bananas, cacao, cassava, coffee, peanuts, pineapples, rice, sugar cane, tobacco; mining—bauxite, clay, gold, gypsum, marble, salt; manufacturing—animal feed, beer, cement, chocolate, glass, molasses, rum, sugar, textiles, vegetable oil
Money: Basic unit—peso
Climate: Tropical wet

The Dominican Republic is a country shaped by both Spanish and African influence. Christopher Columbus landed there on his first voyage to the New World in 1492. The capital, Santo Domingo, was the first European-built city in the Western Hemisphere. The University of Santo Domingo, founded in 1538, is the hemisphere's oldest university.

Most Dominicans speak Spanish and are black or of mixed black and European descent. A small percentage are white. While a majority of the people are Roman Catholics, others practice African voodoo religions. Many apartment houses and other city buildings in the Dominican Republic have Spanish-style architecture. The country's popular music shows both Spanish and African influences.

About two-thirds of all Dominican workers are farmers. Most work on their own small farms. Others work on large plantations. Sugar ranks as the leading crop, and sugar refining is the country's leading industry.

Ecuador

Ecuador (43) is located in South America. It is bordered by Colombia and Peru.

Capital: Quito
Official Name: República del Ecuador (Republic of Ecuador)
Official Language: Spanish
Government: Republic—though formerly under military rule, in 1979, elections were allowed to be held for a new civilian government; people elected a civilian president and Congress
National Anthem: "Himno Nacional del Ecuador" ("National Anthem of Ecuador")
Flag: Top half yellow; bottom half divided equally into two horizontal stripes of blue and red; Ecuador's coat of arms lies in center of flag
Area: 109,484 sq. mi. (283,561 km²); greatest distances—north-south, 450 mi. (724 km); east-west, 395 mi. (636 km); coastline—1,278 mi. (2,057 km), including the Galapagos Islands
Population: 1983 estimate—9,235,000; distribution—57 per cent rural, 43 per cent urban; density—85 persons per sq. mi. (33 persons per km²)
Largest Cities: (1974 census) Guayaquil (814,064) and Quito (597,113)
Economy: Mainly agricultural, with some mining, manufacturing, and trade
Chief Products: Agriculture—bananas, barley, cacao, cattle, coffee, corn, cotton, pyrethrum, rice, sugar, vegetables, wheat; forestry—balsa wood, tagua nuts; manufacturing—building materials, cement, chemicals, drugs, flour, processed foods, hats, leather, textiles; mining—copper, gold, petroleum
Money: Basic unit—sucre; one hundred centavos equal one sucre
Climates: Desert, highlands, tropical wet and dry, tropical wet

Ecuador exports more bananas than any other country, but petroleum ranks as its chief export. Ecuador is a rapidly developing nation. About a sixth of all Ecuadoreans work in manufacturing.

Most Ecuadoreans speak Spanish, and a majority of them are Roman Catholics. About four-fifths of the people are Indians or mestizos, persons of mixed Indian and European ancestry. Whites of European ancestry and blacks each make up about 10 per cent of the population.

Many of the mestizos work on large banana and cacao plantations owned by whites. Others work their own small plots of land. Most of the Indians live in rural villages in the Andes Mountains, making a bare living farming and herding. They speak their own languages, live in small adobe houses, and follow Indian customs. Market days are a traditional time of celebration for the Indians. The villagers gather to trade, meet friends, and enjoy music and dancing.

Egypt

Egypt (44) is located in northeastern Africa. It is bordered by Israel, Sudan, and Libya.

Capital: Cairo
Official Name: Arab Republic of Egypt
Official Language: Arabic
Government: Republic—one-house legislature called People's Assembly nominates one candidate to run for president; president appoints one or more vice-presidents and a Cabinet, which helps plan national policy
National Anthem: "Beladi, Beladi" ("My Country, My Country")
Flag: Three horizontal stripes of red, white, and black; red stands for sacrifice, white for purity, and black for the past; the hawk emblem, which lies in center of flag, was the mark of the tribe of Muhammad, the founder of Islam
Area: 386,662 sq. mi. (1,001,449 km²); greatest distances—east-west, 770 mi. (1,240 km); north-south, 675 mi. (1,086 km); coastline—Mediterranean Sea, 565 mi. (909 km); Red Sea, 850 mi. (1,370 km)
Population: 1983 estimate—44,697,000; distribution—56 per cent rural, 44 per cent urban; density—117 persons per sq. mi. (45 persons per km²)
Largest Cities: (1976 est.) Cairo (6,133,000) and Giza (933,900); (1975 est.) Alexandria (2,320,000)
Economy: Largely agricultural, but with growing industries
Chief Products: Agriculture—beans, clover, corn, cotton, millet, onions, potatoes, rice, sugar cane, wheat; fishing—sardines, shrimps; manufacturing—cement, chemicals, fertilizers, paper, processed foods, steel, textiles; mining—gypsum, iron ore, manganese, petroleum, phosphate rock, salt
Money: Basic unit—pound; one hundred piasters equal one pound
Climates: Steppe, desert

More than half of the Egyptian people live in rural areas, in small and crowded villages. The farmers use the great Nile River to irrigate their fields of cotton and other crops. Their houses are made of sun-dried brick. The whole family goes to the market place once a week, taking butter, chickens, eggs, and vegetables to sell or trade. The other rural people are the Bedouins of the desert, who move about in search of fresh pastureland for their camels, goats, and sheep. Their homes are their tents.

Rich, poor, and a growing middle class live in the cities. All but the poor follow Western ways of life in their clothing, food, housing, and leisure activities.

Agriculture is the chief occupation of the country. Since 1950, however, the government has set up many activities in manufacturing, mining, power production, and transportation and communications.

El Salvador

El Salvador (45) is located in Central America. It is bordered by Honduras and Guatemala.

Capital: San Salvador
Official Name: República de El Salvador (Republic of El Salvador)
Official Language: Spanish
Government: Republic—new government established in 1982 to replace military junta or council; people elected a Constituent Assembly to rewrite country's laws; Assembly appointed a temporary president until election could be held
National Anthem: "Himno Nacional" ("National Hymn")
Flag: Two blue horizontal stripes separated by a white stripe; nation's coat of arms lies in center of white stripe
Area: 8,124 sq. mi. (21,041 km²); greatest distances—north-south, 88 mi. (142 km); east-west, 163 mi. (262 km); coastline—189 mi. (304 km)
Population: 1983 estimate—5,229,000; distribution—61 per cent rural, 39 per cent urban; density—645 persons per sq. mi. (249 persons per km²)
Largest Cities: (1977 est.) San Salvador (397,126), Santa Ana (112,830), and San Miguel (72,874)
Economy: Mainly farming and the export of agricultural products
Chief Products: Agriculture—beans, coffee, corn, cotton, rice, sugar cane; manufacturing—chemicals, cigarettes, processed foods and beverages, leather goods, textiles
Money: Basic unit—colón
Climate: Tropical wet and dry

This small tropical land of cone-shaped volcanoes, green valleys, and scenic lakes has coffee as its leading crop. It also has one of the world's fast-growing populations. There are many social problems largely resulting from the unequal distribution of wealth in El Salvador. Well-to-do landowners make up only 3 per cent of the population but earn about half the nation's annual income. Thousands of poor farmers have moved to the large cities, but the nation's developing industries cannot provide jobs for all the unemployed.

About 92 per cent of all Salvadorans are mestizos, persons of mixed Indian and Spanish descent. Nearly 5 per cent of the people are of unmixed white ancestry, and 3 per cent are Indians. Most are Roman Catholics, and most speak Spanish, the nation's official language.

El Salvador's most colorful religious festival celebrates the Feast of the Holy Saviour of the World. It lasts from July 24 to August 6 and includes carnival rides, fireworks, folk dancing, and processions.

Equatorial Guinea

Equatorial Guinea (46) is located on the west coast of Africa. It is bordered by Cameroon and Gabon.

Capital: Malabo on Fernando Po
Official Name: República de Guinea Ecuatorial (Republic of Equatorial Guinea)
Official Language: Spanish
Government: Military rule—headed by three-member military council; one military officer serves as president and two as vice-presidents; a Cabinet helps carry out government operations
National Anthem: "Caminemos pisando la senda de nuestra inmensa felicidad" ("Let's walk through the jungle of our immense happiness")
Flag: Green, white, and red horizontal stripes, and a blue triangle at the staff; the national coat of arms is on the white stripe
Area: 10,830 sq. mi. (28,051 km^2); greatest distances—east-west, 205 mi. (338 km); north-south, 200 mi. (320 km); coastline—212 mi. (341 km)
Population: 1983 estimate—389,000; distribution—54 per cent urban, 46 per cent rural; density—36 persons per sq. mi. (14 persons per km^2)
Largest Cities: (1973 est.) Malabo (60,000) and Bata (30,000)
Economy: Based on agriculture, forestry, and fishing
Chief Products: Bananas, cocoa, coffee, timber
Money: Basic unit—ekuele
Climate: Tropical wet

Equatorial Guinea was a Spanish province until Spain granted the tiny nation its independence in 1968. The country consists of two provinces: Rio Muni, a stretch of tropical rain forest that lies between Cameroon and Gabon on the west coast of Africa; and Fernando Po, an island that lies in the Gulf of Guinea. Fernando Po's rich, volcanic soil produces valuable banana, coffee, and cacao crops. Rio Muni produces good lumber.

About 75 per cent of the nation's people live in Rio Muni. Most are members of the Fang, a black African ethnic group. The majority of the people of Fernando Po are also black Africans who belong to the Bubi and Fernandino ethnic groups.

Most of the nation's rural people are farmers. Others work in lumber camps or engage in fishing. Most urban people work in small industries or in import-export activities. Although Spanish is the country's official language, Fang is the most widely used language.

Ethiopia

Ethiopia (47) is located in eastern Africa. It is bordered by Sudan, Djibouti, Somalia, and Kenya.

Capital: Addis Ababa
Official Language: Amharic
Government: Military rule—ruled by 120-member military council; council chairman serves as head of state; also has a civilian Cabinet which, in theory, has little control
Flag: Three horizontal stripes—green, yellow, and red (from top to bottom); the colors of the flag may stand for the Holy Trinity, the rainbow, or Ethiopia's three main provinces
Area: 471,778 sq. mi. (1,221,900 km^2); greatest distances—east-west, 1,035 mi. (1,656 km); north-south, 1,020 mi. (1,632 km); coastline—628 mi. (1,011 km)
Population: 1983 estimate—33,552,000; distribution—87 per cent rural, 13 per cent urban; density—70 persons per sq. mi. (27 persons per km^2)
Largest Cities: (1980 est.) Addis Ababa (1,277,159) and Asmara (443,060)
Interesting Sight: Addis Ababa, the capital of Ethiopia and one of Africa's leading cities
Economy: Largely dependent upon agriculture, with some mining but very limited industry
Chief Products: Agriculture—barley, beans, beeswax, coffee, cotton, hides and skins, livestock (cattle, goats, sheep), lumber, millet, oilseeds, peas, sugar, wheat; mining—gold, platinum
Money: Basic unit—birr
Climates: Steppe, desert, tropical wet and dry

Ethiopia is a beautiful country. Rugged mountains, scenic valleys, and ancient churches carved from stone make it a tourist attraction. Elephants, buffaloes, giraffes, leopards, lions, rhinoceroses, zebras, and many other animals abound. Hot, dry desert lies on the northern and southern borders, but most of the country lies on a high plateau that has good soil, valuable mineral deposits, and a cool climate.

Most Ethiopians are farmers who raise crops on small plots of land on the cool, windswept highlands. On the baking-hot lowlands, nomads move from place to place in search of water and pasture for their livestock. The official religion is the Ethiopian Christian Church, but there are many Muslims living in the country. There are small groups of Ethiopian Jews, called *Falashas,* and some people practice traditional African religions. Only about 5 per cent of the people can read and write.

Fiji

Fiji (48) is an island nation in the southwest Pacific Ocean.

Capital: Suva
Official Language: English
Government: Constitutional monarchy—two-house parliament consisting of Senate and House of Representatives; prime minister, country's political leader, is leader of the majority party of the House of Representatives
National Anthem: "God Save the Queen"
Flag: The British Union Jack appears in the upper left on a light blue field; on the right is the shield from Fiji's coat of arms with a British lion, a dove, coconut palms, and such agricultural products as bananas and sugar cane
Area: 7,056 sq. mi. (18,274 km²); greatest distances—north-south, 364 mi. (586 km); east-west, 334 mi. (538 km); coastline—925 mi. (1,489 km)
Population: 1983 estimate—664,000; distribution—63 per cent rural, 37 per cent urban; density—93 persons per sq. mi. (36 persons per km²)
Largest Cities: (1976 census) Suva (63,628) and Lautoka (22,672)
Economy: Based upon agriculture and the export of agricultural products, with some tourism and manufacturing
Chief Products: Agriculture—bananas, coconuts, forest products, sugar; manufacturing—beer, cement, cigarettes; mining—gold, silver
Money: Basic unit—Fijian dollar
Climate: Tropical wet

Fiji is made up of more than 800 scattered islands. There are three major groups of people. Forty per cent are native Fijians chiefly of Melanesian descent. About 50 per cent are descendants of laborers imported from India between 1879 and 1916 to work on Fiji's sugar plantations. The remaining 10 per cent—Fiji's so-called general population group—have Chinese, European, Micronesian, or Polynesian ancestry. Most native Fijians live in rural areas, and most Indians still work in the cane fields; but others have become prosperous shopkeepers or business people.

Fiji became independent in 1970 after having been a British colony since 1874. Its government has a prime minister and two houses of parliament.

English is the official language of Fiji. Children are not required by law to attend school. However, over 85 per cent of those from 6 to 13 years old do so. The University of the South Pacific in Suva serves students from hundreds of the Pacific islands.

Finland

Finland (49) is located in northern Europe. It is bordered by Norway, Russia, and Sweden.

Capital: Helsinki
Official Name: Republic of Finland
Official Languages: Finnish and Swedish
Government: Republic—president, nation's head of state and chief executive, appoints the prime minister who heads the government; prime minister selects a Cabinet to help set government programs; one-house legislature, the *Eduskunta,* is elected by the people
National Anthem: In Finnish, "Maamme"; in Swedish, "Vårt Land" ("Our Land")
Flag: Two blue stripes—one horizontal, one vertical—forming a cross on a white field; Finland's red and yellow coat of arms lies at the center of the cross
Area: 130,129 sq. mi. (337,032 km^2), including 12,206 sq. mi. (31,613 km^2) of inland water; greatest distances—east-west, 320 mi. (515 km); north-south, 640 mi. (1,030 km); coastline—1,462 mi. (2,353 km)
Population: 1983 estimate—4,821,000; distribution—60 per cent urban, 40 per cent rural; density—36 persons per sq. mi. (14 persons per km^2)
Largest Cities: (1977 est.) Helsinki (499,205) and Tampere (165,418)
Economy: Mainly based upon industry, with agriculture, commerce, forestry, and trade also important
Chief Products: Agriculture—barley, cattle, dairy products, eggs, oats, potatoes, rye, sugar beets, wheat; forestry—birch, pine, spruce; manufacturing—chemicals, machinery, metals, paper and pulp, processed foods, textiles and clothing, transportation equipment, wood and wood products; mining—copper, granite, iron, limestone
Money: Basic unit—markka
Climates: Subarctic, continental moist

Finland has historically been fought over, occupied, and partially annexed by nearby countries; but it is now an independent republic. Its economy is based mostly on private ownership, and the Finns have a high standard of living.

A country of thousands of lovely lakes and thick forests, Finland also has a long, deeply indented coast marked by colorful red and gray granite rocks. Most Finns live in the southern part of their country, and about three-fifths live in cities and towns. In the cities, most people own or rent apartments. In the country, most live in one-family homes on farms or in villages. The most famous feature of Finnish life is a special kind of bath called a sauna, a steam bath produced by throwing water on hot stones. Most Finns take a sauna at least once a week.

France

France (50) is located in western
Europe. It is bordered by Belgium,
Luxembourg, West Germany, Switzer-
land, Italy, Monaco, Andorra, and Spain.

Capital: Paris
Official Language: French
Government: Republic—two-house parliamentary form of government; president is elected by the people and appoints
   the prime minister and the Council of Ministers
   or Cabinet; together they direct day-to-day government operations
National Anthem: "La Marseillaise"
Flag: Three vertical stripes of blue, white, and red (left to right)
Area: Metropolitan France (mainland and Corsica)—211,208 sq. mi. (547,026
   km^2); greatest distances—east-west, 605 mi. (974 km); north-south, 590 mi.
   (950 km); coastline—2,300 mi. (3,701 km)
Population: Metropolitan France, 1983 estimate—54,360,000; distribution—78
   per cent urban, 22 per cent rural; density—256 persons per sq. mi. (99
   persons per km^2)
Largest Cities: (1975 census) Paris (2,299,830), Marseille (908,600), and Lyon
   (456,716)
Economy: Mainly industrial, with some agriculture
Chief Products: Agriculture—barley, corn, flowers, flax, fruits, livestock, oats,
   potatoes, rice, rye, sugar beets, wheat; fishing—cod, crabs, herring, lobsters,
   mackerel, oysters, sardines, shrimps, tuna; manufacturing—aircraft, aluminum, automobiles, chemicals, clothing, dairy products, electrical and non-
   electrical machinery, furniture, iron and steel, jewelry, paper, perfume, textiles, wine; mining—bauxite, coal, gypsum, iron ore, potash, uranium
Money: Basic unit—franc; one hundred centimes equal one franc
Climates: Oceanic moist, subtropical dry summer

France is not only a beautiful and historic country, it is also rich and
powerful. France plays an important part in world politics. It has great
automobile, chemical, and steel industries. As measured by exports,
France stands fifth among the countries of the world in its trade with
other nations. The country's most important natural resource is its rich
soils. It is a leader in growing wheat, vegetables, and many other crops.

The French people are famous for their enjoyment of living. Good
food and good wine are a part of everyday life. French artists and writers
have excelled in literature, painting, and sculpture and have set an example of excellence around the world. French architecture has produced
Gothic cathedrals, French châteaux, and modern apartment buildings.
The city of Paris is a world capital of art and learning.

Gabon

Gabon (51) is located in west-central Africa. It is bordered by Equatorial Guinea, Cameroon, and the Congo.

Capital: Libreville
Official Name: République Gabonaise (Gabon Republic)
Official Language: French
Government: Republic—president, elected by people, serves as head of state and chief administrator; president appoints a council of ministers to help govern the country; one-house legislative body is also elected by the people
Flag: Three horizontal stripes of green, yellow, and blue (top to bottom)
Area: 103,347 sq. mi. (267,667 km^2); greatest distances—east-west, 400 mi. (640 km); north-south, 430 mi. (688 km); coastline—500 mi. (800 km)
Population: 1983 estimate—569,000; distribution—64 per cent rural, 36 per cent urban; density—5 persons per sq. mi. (2 persons per km^2)
Largest Cities: (1976 est.) Libreville (251,000) and Port-Gentil (85,000)
Economy: Mainly agriculture and forestry, but with a growing mining industry
Chief Products: Agriculture—bananas, cacao, cassava, coffee, yams; fishing—deep sea, fresh water, whaling; forestry—mahogany, okoumé; mining—gold, iron ore, manganese, petroleum, uranium
Money: Basic unit—franc
Climates: Tropical wet and dry, tropical wet

The republic of Gabon is a small, heavily forested country straddling the equator. It was here that Albert Schweitzer, the much-honored physician, missionary, and musician, built his hospital and leper house. The site is near Lambaréné, an inland town.

Gabon has high-quality lumber and rich mineral deposits, particularly iron and manganese. There are many ethnic groups. The Fang, the most important, live in the north. Once fierce warriors, they were feared by neighboring groups and by Europeans. They now dominate the national government. A related group, the Omyéné, live along the coast. Isolated from other people, small groups of Pygmies live in the thick southern forest. Many Gabonese, particularly those living in the towns, are Christians. Others follow traditional African religions. Music and dancing play an important part in their ceremonies.

Most adults in Gabon cannot read and write. About 90 per cent of the children now go to primary schools.

Gambia

Gambia (52) is located in western Africa. It is bordered by Senegal.

Capital: Banjul
Official Name: The Gambia
Official Language: English
Government: Republic—controlled by a president and a Cabinet; president selects Cabinet ministers from members of Parliament
Flag: Three horizontal bands of red, blue, and green (top to bottom) divided by two narrow white bands
Area: 4,361 sq. mi. (11,295 km²); greatest distances—east-west, 205 mi. (328 km); north-south, 35 mi. (56 km); coastline—44 mi. (71 km)
Population: 1983 estimate—653,000; distribution—84 per cent rural, 16 per cent urban; density—150 persons per sq. mi. (58 persons per km²)
Largest City: (1978 est.) Banjul (45,600)
Economy: Mainly agricultural
Chief Products: Agriculture—bananas, cassava, corn, hides and skins, limes, livestock (cattle, goats, sheep), mangoes, millet, oranges, palm kernels, papayas, peanuts, rice, vegetables
Money: Basic unit—dalasi
Climate: Steppe

Once a slave-trading center, the tiny nation of Gambia is one of the poor independent African nations. Peanuts are its main crop, and peanut processing its only industry.

Almost all of the people are black Africans. The Malinke people live throughout Gambia. They are a tall, music-loving people who make a meager living as traders and peanut farmers. Most of the Fulani live in eastern Gambia and raise cattle for a living. The Wolof people are farmers near Gambia's northern border. They enjoy dancing and music. The women often dress elegantly in turbans and full-skirted dresses. The Jola, south of the Gambia River near the coast, are hard-working farmers who live in small villages surrounded by earthen walls. They raise rice and millet for food. In eastern Gambia, where the soil is poor and farming is difficult, live the Seraculeh. Every year people from Senegal called the strange farmers come to help the Seraculeh plant and harvest the crops. In exchange for their help, the strange farmers take an agreed share of the crop and are also given a plot of land on which they raise crops of their own.

Germany, East

East Germany (53) is located in north-central Europe. It is bordered by Poland, Czechoslovakia, and West Germany.

Capital: East Berlin
Official Name: Deutsche Demokratische Republik (German Democratic Republic)
Official Language: German
Government: Communist dictatorship—one-house parliament called *Volkskammer* or People's Chamber which follows Communist leaders' decisions when passing laws or electing members of government bodies
National Anthem: "Auferstanden aus Ruinen" ("Arisen from Ruins")
Flag: Three horizontal stripes of black, red, and gold (top to bottom); nation's coat of arms lies in center of flag
Area: 41,768 sq. mi. (108,178 km^2) including East Berlin; greatest distances—north-south, 315 mi. (507 km); east-west, 225 mi. (362 km); coastline—220 mi. (354 km)
Population: 1983 estimate—16,637,000, including East Berlin; distribution—76 per cent urban, 24 per cent rural; density—399 persons per sq. mi. (154 persons per km^2)
Largest Cities: (1978 est.) East Berlin (1,128,983) and Leipzig (563,980)
Economy: Based largely on industry and services
Chief Products: Agriculture—barley, dairy products, livestock, oats, potatoes, rye, sugar beets, wheat; fishing—cod, herring; manufacturing—chemicals, clothing, electrical equipment, iron and steel, machinery, optical goods, processed foods and metals, textiles, transportation equipment; mining—copper, iron ore, lignite, potash, rock salt, tin
Money: Basic unit—Deutsche mark
Climate: Continental moist

East Germany is the eastern portion of a divided Germany. During World War II, the Russians occupied this part of Germany and, following the war, created there a Communist political and economic system.

The government of East Germany owns more than 85 per cent of the total means of production. The economy has grown rapidly since World War II, though its recovery has not been so great as West Germany's. A major reason was the flow of refugees from East Germany to West Germany, and East Germany's consequent loss of skilled workers. In 1961, the Communists built the Berlin Wall to cut off the refugees' access route.

Industry is the main economic activity in East Germany, but agriculture represents a large chunk. The government owns about 93 per cent of the farmland. See also **Germany, West**.

61

Germany, West

West Germany (54) is located in north-central Europe. It is bordered by Denmark, The Netherlands, East Germany, Czechoslovakia, Austria, Switzerland, France, Belgium, and Luxembourg.

Capital: Bonn
Official Name: Bundesrepublik Deutschland (Federal Republic of Germany)
Official Language: German
Government: Federal republic—two-house parliamentary form of government
National Anthem: Third stanza of "Deutschland-Lied" ("Song of Germany")
Flag: Three horizontal stripes of black, red, and gold (top to bottom)
Area: 96,005 sq. mi. (248,651 km^2), including West Berlin; greatest distances—north-south, 540 mi. (869 km); east-west, 360 mi. (579 km); coastline—354 mi. (570 km)
Population: 1983 estimate—61,412,000, including West Berlin; distribution—85 per cent urban, 15 per cent rural; density—640 persons per sq. mi. (247 persons per km^2)
Largest Cities: (1978 est.) West Berlin (1,909,706) and Hamburg (1,664,305)
Economy: Mainly industrial, with trade, finance, and agriculture also important
Chief Products: Agriculture—barley, dairy products, fruits, livestock, potatoes, rye, sugar beets, wheat; fishing—cod, herring, redfish, shrimp; manufacturing—automobiles, chemicals, clothing, cement, electrical equipment, iron and steel, machinery, processed foods and metals, textiles; mining—coal, iron ore, lead, petroleum, potash, rock salt, zinc
Money: Basic unit—Deutsche mark
Climates: Continental moist, oceanic moist

West Germany grew out of the German zone occupied by three of the Western Allies (United States, Great Britain, and France) following World War II. The government was set up as a democratic republic and remains so. This nation has a parliament composed of two houses: the *Bundestag* (Federal Diet) and the *Bundesrat* (Federal Council). The Bundestag elects a member of the strongest political party in that house to be federal chancellor, the head of government.

The Germans are proud of their heritage. They are famous for being thrifty, hard-working, and obedient to authority. They have striven to regain national prosperity following the devastation of the war.

Germans love music, dancing, beer, good food, and fellowship. German writers, musicians, artists, and architects have made great contributions to world culture. German philosophers especially have had great influence on Western thought. See also **Germany, East**.

Ghana

Ghana (55) is located in western Africa. It is bordered by Upper Volta, Togo, and Ivory Coast.

Capital: Accra
Formerly Called: Gold Coast
Official Language: English
Government: Military rule—the military leader heads the government and serves as chairman of the Provisional National Defense Council, Ghana's governing body
Flag: Horizontal red, yellow, and green stripes with a black star symbolizing African freedom in the center
Area: 92,100 sq. mi. (238,537 km^2); greatest distances—north-south, 445 mi. (716 km); east-west, 310 mi. (499 km); coastline—335 mi. (539 km)
Population: 1983 estimate—12,585,000; distribution—69 per cent rural, 31 per cent urban; density—137 persons per sq. mi. (53 persons per km^2)
Largest Cities: (1970 census) Accra (633,880), Kumasi (345,117), and Sekondi-Takoradi (160,868)
Economy: Mainly agricultural, with some mining
Chief Products: Agriculture—cacao, coffee, copra, kola nuts, palm kernels; mining—bauxite, diamonds, gold, manganese; forestry—mahogany, sapele, utile
Money: Basic unit—new cedi
Climates: Steppe, tropical wet and dry, tropical wet

Ghana is a tropical country and most of its people are black African farmers. Before gaining its independence, Ghana was a British colony called the Gold Coast. In 1960, the nation voted to become a republic, and elected Kwame Nkrumah president. Six years later, a military council seized the government, and since then the nation has gone back and forth between civil and military rule.

English is the official language, but many Ghanaians speak African languages. The national dress in Ghana is made from brightly colored *kente* cloth. Men wrap the cloth around them; women make blouses and narrow skirts from it. Many Ghanaians now also wear Western-style clothing.

About 60 per cent of the people practice traditional African religions. Roughly 30 per cent are Christians, with 10 per cent Muslims. Children in Ghana must attend school until they are 12 years old.

Great Britain

Great Britain (56) is an island nation in northwestern Europe. It is really four countries united under one government: England, Northern Ireland, Scotland, and Wales. It is bordered by the Republic of Ireland.

Capital: London
Official Name: The United Kingdom of Great Britain and Northern Ireland
Also Called: United Kingdom; Britain
Official Language: English
Government: Constitutional monarchy—monarch reigns, but does not rule; cabinet of ministers actually rules the country; the ministers are responsible to Parliament, the nation's two-house, lawmaking body
National Anthem: "God Save the Queen" (or "King")
Flag: Blue background with large red cross outlined in white; red and white stripes form rays from the center of the cross to each corner of the flag
Area: 94,249 sq. mi. (244,104 km^2); greatest distances—north-south, about 600 mi. (970 km); east-west, about 300 mi. (480 km); coastline—2,521 mi. (4,057 km)
Population: 1983 estimate—55,705,000; distribution—91 per cent urban, 9 per cent rural; density—591 persons per sq. mi. (228 persons per km^2)
Largest Cities: (1981 census) Greater London (6,696,008), Birmingham (1,006,908), and Glasgow (762,288)
Economy: Mainly manufacturing and services
Chief Products: Agriculture—barley, beef cattle, oats, pigs, potatoes, poultry and eggs, sheep, sugar beets, wheat; manufacturing—chemicals, clothing, electrical goods, foods and beverages, glass and glassware, machinery, metal goods, motor vehicles, paper, ships, steel, textiles; fishing industry—cod, haddock, herring, plaice, shellfish
Money: Basic unit—pound
Climate: Oceanic moist

Great Britain is small, has few natural resources, and contains less than 2 per cent of the world's people. Yet for hundreds of years it has been one of the world's important countries.

The British started the Industrial Revolution. They founded the largest empire in history, and they took their British ways around the globe. Many of the world's legislatures have borrowed features of the British system of government.

Great Britain gave up its empire, and it has faced many economic and social problems since World War II. It has sought to provide every British citizen with health and welfare "from the cradle to the grave."

Greece

Greece (57) is located in southeastern Europe. It is bordered by Bulgaria, Yugoslavia, Albania, and Turkey.

Capital: Athens
Official Name: Elliniki Dimokratia (Hellenic Republic)
Official Language: Greek
Government: Republic—president serves as head of state and is elected by one-house Parliament; prime minister is head of government and is appointed by president
National Anthem: "Imnos pros tin Eleftherian" ("The Hymn to Liberty")
Flag: A white cross on a blue background in upper left corner; five horizontal blue stripes represent the sea and sky; four horizontal white stripes represent the purity of the Greek struggle for independence; the white cross symbolizes the Greek Orthodox religion
Area: 50,944 sq. mi. (131,944 km²); greatest mainland distances—north-south, 365 mi. (587 km); east-west, 345 mi. (555 km); coastline (including islands)—9,333 mi. (15,020 km)
Population: 1983 estimate—9,919,000; distribution—62 per cent urban, 38 per cent rural; density—194 persons per sq. mi. (75 persons per km²)
Largest Cities: (1971 census) Athens (867,023), Salonika (345,799), and Piraeus (187,362)
Interesting Sights: The Acropolis of Athens; Temple of Apollo at Delphi
Economy: Based mainly on industry and services, with agriculture also contributing
Chief Products: Agriculture—cotton, goats, grapes, lemons, olives, sheep, tobacco, vegetables, wheat; manufacturing—cigarettes, clothing, processed foods and beverages, textiles; mining—bauxite, lignite, chromite
Money: Basic unit—drachma; one hundred lepta equal one drachma
Climate: Steppe

The ancient Greeks were the first people to develop a democratic way of life. The physical remains of their great civilization can be seen today in imposing and beautiful ruins such as the Parthenon and other temples in Athens. Greece is one of the less developed European countries today, and its people have low incomes.

Greeks are lively and take great delight in conversation and company. A majority of Greeks live in farm towns, most of which have a central square where the people meet to talk. Greek Orthodox religious festivals are an important part of life, although there is freedom of worship.

In recent years Greece has abolished its monarchy and alternated between military and civilian governments. In 1974 the democratic civilian forces were voted in, and they held their majority in elections in 1977.

Grenada

Grenada (58) is an island nation located in the Windward Islands of the West Indies.

Capital: St. George's
Official Name: State of Grenada
Official Language: English
Government: Constitutional monarchy—rebels presently rule the country and have suspended the constitution; previously, the country had a two-house legislature with a prime minister serving as chief executive
National Anthem: "The Grenada National Anthem"
Flag: The red-bordered flag has a yellow triangle at the top and bottom and a green triangle on each side; a yellow and brown nutmeg represents Grenada's chief product, and seven gold stars symbolize its seven parishes or districts
Area: 133 sq. mi. (344 km²); greatest distances—north-south, 21 mi. (34 km); east-west, 12 mi. (19 km); coastline—75 mi. (121 km)
Population: 1983 estimate—115,000; density—865 persons per sq. mi. (334 persons per km²)
Largest City: (1976 est.) St. George's (10,000)
Interesting Sights: Grand Etang, a lake in the crater of a volcano, located in the center of Grenada; St. George's, the country's scenic capital city
Economy: Mainly agriculture, fishing, and tourism
Chief Products: Agriculture—bananas, cocoa, mace, nutmeg; manufacturing—food products, beer, rum
Money: Basic unit—East Caribbean dollar
Climate: Tropical wet

Grenada, with its pleasant climate and beautiful scenery and beaches, attracts many tourists. The nation is also one of the world's leading producers of nutmeg and other spices.

In 1974, Grenada gained its independence from Great Britain to become a constitutional monarchy and a member of the Commonwealth of Nations. In 1979, rebels overthrew the government, suspended the constitution, and now rule the country.

About 95 per cent of Grenada's people have African or mixed African and European ancestry. Descendants of East Indians or of Europeans make up the rest of the population. Most Grenadians speak English or one of its dialects. The people of some regions speak a French dialect. More than half the population are Roman Catholics.

Over 30,000 students are enrolled in elementary or high school. Attendance is not required by law.

66

Guatemala

Guatemala (59) is located in Central America. It is bordered by Mexico, Honduras, El Salvador, and Belize.

Capital: Guatemala City
Official Language: Spanish
Government: Military rule—in 1982, a military junta or council took control of the government; later that year, the junta's leader removed the other junta members from power and declared himself the country's only leader
National Anthem: "Himno Nacional de Guatemala" ("National Anthem of Guatemala")
Flag: Two blue vertical stripes representing the Atlantic and Pacific oceans, which form part of the country's borders, separated by a white stripe; Guatemala's coat of arms lies in center of white stripe
Area: 42,042 sq. mi. (108,889 km²); greatest distances—north-south, 283 mi. (455 km); east-west, 261 mi. (420 km); coastlines—Pacific, 152 mi. (245 km); Caribbean, 53 mi. (85 km)
Population: 1983 estimate—7,935,000; distribution—64 per cent rural, 36 per cent urban; density—189 persons per sq. mi. (73 persons per km²)
Largest Cities: (1973 census) Guatemala City (700,504) and Quezaltenango (53,021)
Economy: Mainly agricultural, with trade and industry also significant
Chief Products: Agriculture—bananas, beans, beef cattle, coffee, corn, cotton, rice, sugar cane, wheat; manufacturing—clothing and textiles, handicrafts, processed foods and beverages
Money: Basic unit—quetzal
Climates: Tropical wet and dry, tropical wet

Almost half the people in tropical Guatemala are Indians whose way of life differs greatly from that of other Guatemalans. Their ancestors, the Maya Indians, built a highly developed civilization hundreds of years before Christopher Columbus landed in America. Today, the Indians live in peasant or farm communities apart from the main life of the country. Most speak Indian languages and wear Indian clothing.

The majority of the other Guatemalans—called Ladinos—are of mixed Indian and Spanish ancestry. They speak Spanish and follow a Guatemalan form of Spanish-American customs. The Ladinos include peasants, laborers, and people in the cities and towns who control government and economy.

Guatemala is a developing country. Its major natural resource is fertile soil. There are mineral deposits, and the many mountain streams are a source of cheap hydroelectric power.

Guinea

Guinea (60) is located in western Africa. It is bordered by Guinea-Bissau, Senegal, Mali, Ivory Coast, Liberia, and Sierra Leone.

Capital: Conakry
Official Name: République de Guinée (Republic of Guinea)
Also Called: The Switzerland of Africa
Official Language: French
Government: Republic—elected president serves as head of state and governs the country aided by Council of Ministers; members of one-house legislature, the National Assembly, are elected by the people
Flag: Three vertical stripes: red (for the spirit of sacrifice), gold (for sun and wealth), and green (for the forests)
Area: 94,926 sq. mi. (245,857 km^2); greatest distances—east-west, 495 mi. (792 km); north-south, 375 mi. (600 km); coastline—219 mi. (352 km)
Population: 1983 estimate—5,415,000; distribution—81 per cent rural, 19 per cent urban; density—57 persons per sq. mi. (23 persons per km^2)
Largest Cities: (1972 est.) Conakry (45,304—metro area); (1967 est.) Kankan (50,000)
Economy: Based mainly upon agriculture and mining, with some commerce
Chief Products: Agriculture—bananas, cassava, citrus fruits, coffee, corn, livestock, millet, palm oil and kernels, peanuts, pineapples, rice, sweet potatoes, taro; mining—bauxite, diamonds, gold, iron ore; manufacturing and processing—aluminum
Money: Basic unit—syli
Climate: Tropical wet and dry

Guinea is a beautiful country lying on the western "bulge" of Africa. From the mangrove swamps along the coast, the land rises to a scenic tableland with rivers, deep valleys, and waterfalls. Inland is green savanna, or grassy plains, and forest-covered mountains. There are many bird sanctuaries and several game reserves.

Guinea was a French colony from the late 1800's until it became independent in 1958. It is a republic in which the people elect a president. A council of appointed ministers and an elected National Assembly are the other two governing bodies at the national level.

There are seven major tribal groups in Guinea and many smaller ones. French is the official language. Tribe members, however, use their own languages more often than French. Most Guineans are farmers and live in small villages. Their houses have mud walls and thatched roofs. They wear Guinea's traditional dress: long, flowing robes.

Guinea-Bissau

Guinea-Bissau (61) is located on the west coast of Africa. It is bordered by Senegal and Guinea.

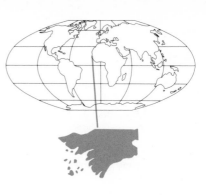

Capital: Bissau
Official Name: Republic of Guinea-Bissau
Official Language: Portuguese
Government: Military rule—nine-member council of six military leaders and three civilians controls the country; head of the council serves as nation's president; the military abolished the National Popular Assembly, Guinea-Bissau's lawmaking body
Flag: A black star is centered on a red vertical stripe to the left of two horizontal stripes; the top horizontal stripe is yellow, and the bottom one is green
Area: 13,948 sq. mi. (36,125 km^2); greatest distances—north-south, 120 mi. (193 km); east-west, 200 mi. (322 km); coastline—247 mi. (398 km)
Population: 1983 estimate—834,000; distribution—76 per cent rural, 24 per cent urban; density—60 persons per sq. mi. (23 persons per km^2)
Largest City: (1971 est.) Bissau (65,000)
Economy: Based largely upon agriculture
Chief Products: Palm kernels, peanuts, rice
Money: Basic unit—escudo; one hundred centavos equal one escudo
Climate: Tropical wet and dry

Peanuts are the leading crop and main export of this country, which is next to Guinea on the African "bulge." The country was a colony of Portugal, which had used the area as a base for slave trade from the 1600's to the 1800's. Between 1968 and 1974, the people fought their war of independence against Portuguese rule. The country declared itself a republic in 1973. In 1980, however, military leaders overthrew the nation's civilian government and now rule the country.

Black Africans make up about 85 per cent of the people. Most of the rest consist of mestizos, or persons of mixed black African and Portuguese ancestry. Most people live in rural areas and make a bare living farming. Many of them live in straw huts with thatched roofs. A majority of the people practice animism, the belief that everything in nature has a spirit. Many other people are Muslims. The official language of Guinea-Bissau is Portuguese. Most people, however, use crioulo, a local language that combines local African languages and Portuguese.

Guyana

Guyana (62) is located on the northeast coast of South America. It is bordered by Suriname, Brazil, and Venezuela.

Capital: Georgetown
Official Name: Cooperative Republic of Guyana
Official Language: English
Government: Republic—president and 53-member legislature called the National Assembly are elected by the people; president appoints a prime minister who chooses a Cabinet; prime minister and Cabinet carry out governmental operations; the National Assembly makes the country's laws
National Anthem: "Guyana National Anthem"
Flag: Green (representing agriculture and forests) with a red triangle (zeal in nation building) and a golden yellow arrowhead (minerals); triangle has a black border (endurance) and the arrowhead has a white border (water resources)
Area: 83,000 sq. mi. (214,969 km^2); greatest distances—north-south, 495 mi. (797 km); east-west, 290 mi. (467 km); coastline—270 mi. (435 km)
Population: 1983 estimate—919,000; distribution—70 per cent rural, 30 per cent urban; density—10 persons per sq. mi. (4 persons per km^2)
Largest Cities: (1976 est.) Georgetown (72,049) and New Amsterdam (17,782)
Economy: Based on agriculture and mining
Chief Products: Agriculture—sugar cane, rice; manufacturing and processing—sugar, rice, timber, coconuts; mining—bauxite, diamonds, gold, manganese
Money: Basic unit—Guyana dollar
Climates: Tropical wet and dry, tropical wet

Guyana is a tropical land. About half its people are East Indians, whose ancestors were brought from India to work on sugar plantations. Most of them live in rural areas, some on the sugar plantations and others on small farms where they grow rice and vegetables. Still others have moved to cities and towns, where they work as merchants, doctors, and lawyers.

About 4 out of 10 persons are blacks whose ancestors were brought from Africa as slaves. Most of these people live in cities and towns. They work as teachers, police officers, government employees, and as skilled workers in the sugar-grinding mills and bauxite mines. The rest of the people are Europeans, Chinese, and Amerindians, or American Indians.

After much racial trouble between Indians and blacks, Guyana saw few outbreaks of violence in the late 1960's. Its economy expanded, and it worked for economic cooperation among Caribbean nations.

Haiti

Haiti (63) is located in the West Indies. It covers the western part of the island of Hispaniola, which lies between Cuba and Puerto Rico in the Caribbean Sea.

Capital: Port-au-Prince
Official Name: République d'Haïti (Republic of Haiti)
Official Language: French
Government: Republic (dictatorship)—head of state rules as a dictator, and controls armed forces and a secret police force; one-house legislature meets twice a year; country's constitution has never been fully enforced
National Anthem: "La Dessalinienne"
Flag: The black left half stands for the blacks of Haiti; the red half represents its mulattoes; in the center is the Haitian coat of arms
Area: 10,714 sq. mi. (27,750 km^2); greatest distances—east-west, 180 mi. (290 km); north-south, 135 mi. (217 km); coastline—672 mi. (1,081 km), including Gonâve and other offshore islands
Population: 1983 estimate—5,284,000; distribution—76 per cent rural, 24 per cent urban; density—492 persons per sq. mi. (190 persons per km^2)
Largest City: (1978 est.) Port-au-Prince (745,700)
Interesting Sight: The National Palace in Port-au-Prince
Economy: Based on agriculture and tourism
Chief Products: Agriculture—coffee, sisal, sugar cane; mining—copper, bauxite
Money: Basic unit—gourde
Climate: Tropical wet

Haiti lies close to Cuba. The nation shares the island of Hispaniola with the Dominican Republic. Haiti has been independent since it threw off French rule in 1804. Most of the time since then, it has been ruled by dictators disinterested in the welfare of the people. One of the less developed countries in the hemisphere, Haiti consists of a population of black farmers who raise barely enough food to feed their families.

Haitians practice a religion that is called voodoo, a blend of Christian and African beliefs. The members dance at certain ceremonies in order to be possessed by gods of rain, love, war, and farming.

About 5 per cent of the people of Haiti are mulattoes, or people of mixed black and white ancestry. Most of the country's mulattoes belong to the middle and upper classes, and many have been educated in France. Most of them live comfortably in modern houses and are prosperous merchants, doctors, and lawyers.

Honduras

Honduras (64) is located in Central America. It is bordered by Nicaragua, El Salvador, and Guatemala.

Capital: Tegucigalpa
Official Language: Spanish
Government: Republic—people elect a president to head the government and a legislature to make the nation's laws; president appoints a Cabinet; military leaders have often taken complete control of government; military leaders are completely responsible for national security policies
National Anthem: "Himno Nacional de Honduras" ("National Hymn of Honduras")
Flag: Two blue horizontal stripes separated by a white stripe; five blue stars lie in center of white stripe
Area: 43,277 sq. mi. (112,088 km^2); greatest distances—east-west, 405 mi. (652 km); north-south, 240 mi. (386 km); coastlines—Caribbean Sea, 382 mi. (615 km); Pacific Ocean, 48 mi. (77 km)
Population: 1983 estimate—4,104,000; distribution—64 per cent rural, 36 per cent urban; density—96 persons per sq. mi. (37 persons per km^2)
Largest City: (1974 census) Tegucigalpa (267,754)
Economy: Mainly agriculture, with some mining and limited manufacturing
Chief Products: Agriculture—bananas, beans, beef and dairy cattle, coffee, corn, cotton, rice, sugar cane, tobacco; manufacturing—clothing and textiles, processed foods and beverages, lumber; mining—silver
Money: Basic unit—lempira; one hundred centavos equal one lempira
Climates: Tropical wet and dry, tropical wet

Many different groups of peoples live in Honduras, but most of them are mestizos, or people with both Spanish and Indian ancestry. Honduras was a Spanish colony until 1821. Since then various factions have run the country, and it has been under the influence of outside interests. There has been frequent political violence, and Honduras has had a military government for many years.

Sixty per cent of the Hondurans live in rural towns or villages. Most of the rural population are poor peasants who own or rent small farms. Modernization is taking place in the cities because of expanding industry and education, but changes are only slowly reaching farms. About 30 per cent of the people cannot read and write.

In the northern lowlands, Hondurans grow bananas, the leading source of income. In the inland mountains, farmers raise beans, cattle, coffee, corn, and tobacco.

Hungary

Hungary (65) is located in central Europe. It is bordered by Czechoslovakia, Russia, Romania, Yugoslavia, and Austria.

Capital: Budapest
Official Name: Magyar Népköztársaság (Hungarian People's Republic)
Official Language: Magyar (Hungarian)
Government: People's republic (Communist dictatorship)—Communist Party controls the government; people elect members of the nation's one-house legislature, called the National Assembly, which elects a Presidential Council from among its members; the chairman of this council is Hungary's head of state; an appointed Council of Ministers heads the various government departments; the chairman of the Council of Ministers serves as Hungary's head of government, or premier
National Anthem: "Himnusz" ("Hymn")
Flag: Three horizontal stripes of red, white, and green (top to bottom)
Area: 35,919 sq. mi. (93,030 km^2); greatest distances—east-west, 312 mi. (502 km); north-south, 193 mi. (311 km)
Population: 1983 estimate—10,818,000; distribution—52 per cent urban, 48 per cent rural; density—300 persons per sq. mi. (116 persons per km^2)
Largest Cities: (1979 est.) Budapest (2,093,187), Miskolc (210,948), Debrecen (199,742), and Szeged (177,677)
Economy: Mainly industry, with some agriculture
Chief Products: Agriculture—corn, dairy products, livestock, potatoes, sugar beets, wheat, wine grapes; manufacturing—alumina, chemicals, foods and beverages, machinery, steel, textiles, transportation equipment; mining—bauxite
Money: Basic unit—forint
Climate: Continental moist

Landlocked Hungary has become industrialized since the late 1940's. Many rural customs are disappearing as modern city ways of life have become popular. But Hungarians still love highly seasoned food, especially with paprika; excellent wine; and their famous, lively folk music.

Soccer is the popular sport, but other favorites are basketball, fencing, and volleyball. Almost all Hungarians read and write. Religious activities are supervised by the state. Most people are Roman Catholics.

Hungary's constitution, which was adopted in 1949, calls the country a people's republic. The Communist Party—the only political party allowed—approves all laws before they go to the legislature, which then formally enacts the laws.

Iceland

Iceland (66) is an island nation located in the North Atlantic Ocean.

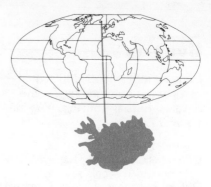

Capital: Reykjavík
Official Name: Lýdhveldidh Ísland
(Republic of Iceland)
Also Called: Land of Frost and Fire
Official Language: Icelandic
Government: Republic—elected president serves as head of state but has little power; prime minister and Cabinet, appointed by president, propose and carry out government policies; a one-house parliament, called the Althing, passes the country's laws
Flag: A red cross edged in white appears on a blue field; blue is the national color, and red and white recall the flag of Denmark, former ruler of Iceland
Area: 39,769 sq. mi. (103,000 km^2); greatest distances—east-west, 300 mi. (483 km); north-south, 190 mi. (306 km); coastline—1,243 mi. (2,000 km)
Population: 1983 estimate—236,000; distribution—87 per cent urban, 13 per cent rural; density—5 persons per sq. mi. (2 persons per km^2)
Largest Cities: (1979 est.) Reykjavík (83,536) and Kopavogur (13,533)
Economy: Mainly fishing and fish processing, with some agriculture and manufacturing
Chief Products: Agriculture—sheep, cattle, hay, market gardening; fishing—cod, herring, haddock; manufacturing and processing—food processing, metal products, clothing, woodworking, printing and bookbinding
Money: Basic unit—krona
Climate: Polar

Iceland is sometimes called the Land of Frost and Fire because large glaciers lie next to steaming hot springs, geysers, and volcanoes. Part of the Gulf Stream flows around Iceland, warming the lowlands and keeping ports free of ice all year long.

Icelanders are descendants of the Norwegians who settled Iceland more than a thousand years ago. Most of the people live in coastal towns and earn their living from the sea. Almost all the country's exports are fish or fish products.

The government is a republic with a president who is elected by the people, a prime minister, the Cabinet, and parliament.

Icelanders like sports. Many people play chess and bridge and are fond of modern and ancient poetry and prose. Reykjavík, the capital, has two theaters and a symphony orchestra. All grammar school students learn at least five foreign languages.

India

India (67) is a large nation located in southern Asia. It is bordered by China, Nepal, Bhutan, Bangladesh, Burma, and Pakistan.

Capital: New Delhi
Official Name: Bharat (Union of India)
Official Language: Hindi
Government: Federal republic—two-house parliament-cabinet form of government; the prime minister, appointed by the president, chooses a cabinet and heads the government with the support of parliament
National Anthem: "Jana-gana-mana" ("Thou Art the Ruler of the Minds of All People"); National Song: "Vande Mataram" ("I Bow to Thee, Mother")
Flag: Three horizontal stripes; saffron on the top, white, and green; the Dharma Chakra, or Wheel of Law, lies in the center of the white stripe
Area: 1,269,346 sq. mi. (3,287,590 km^2); greatest distances—north-south, about 2,000 mi. (3,200 km); east-west, about 1,700 mi. (2,740 km); coastline—4,252 mi. (6,843 km), including 815 mi. (1,312 km) of coastline of island territories
Population: 1983 estimate—726,154,000; distribution—78 per cent rural, 22 per cent urban; density—572 persons per sq. mi. (221 persons per km^2)
Largest Cities: (1971 census) Bombay (5,970,575) and Delhi (3,287,883)
Economy: Mainly agriculture, with some mining and manufacturing
Chief Products: Agriculture—cotton, jute, peanuts, pepper, rice, sugar cane, tea, tobacco, vegetables, wheat; manufacturing and processing—brassware and silverware, cotton and silk materials, fertilizer, iron and steel, jute bags and rope, leather goods, paper, rugs, sugar, woodwork; mining—bauxite, coal, iron ore, manganese ore, mica, salt
Money: Basic unit—rupee
Climates: Steppe, desert, highlands, tropical wet and dry, tropical wet

India has more people than any other country except China. They belong to many different races and religions and speak about 180 languages. Some Indians have great wealth, but others sleep in the streets and live on only a few cents a day. There are some college graduates but numerous others have never gone to school at all.

Ancient customs may be seen side by side with the latest advances of civilization and science. Cows, which India's millions of Hindus consider sacred, often roam freely in modern business districts.

One of the famous world leaders was Mohandas K. Gandhi, who, through nonviolent programs, led the people of India to independence. Now India has a parliamentary form of government, and its constitution includes many parts drawn from the U.S. Constitution.

Indonesia

Indonesia (68) is an island nation
in Southeast Asia. It is bordered
by Malaysia and Papua New Guinea.

Capital: Jakarta
Official Name: Republic of Indonesia
Official Language: Indonesian
Government: Republic—president heads
  government; in theory, the president
  and a People's Consultative Assembly are sup-
  posed to establish government policies; in prac-
  tice, the president, with top army leaders, makes all important decisions; the
  People's Representative Council, the nation's parliament, passes laws based
  on the decisions
National Anthem: "Indonesia Raya" ("Greater Indonesia")
Flag: Top half is red, which stands for courage; bottom half is white, which
  represents purity
Area: 788,425 sq. mi. (2,042,012 km^2); greatest distances—east-west, about
  3,200 mi. (5,150 km); north-south, about 1,200 mi. (1,930 km); coastline—
  22,888 mi. (36,835 km)
Population: 1983 estimate—156,864,000; distribution—80 per cent rural, 20
  per cent urban; density—199 persons per sq. mi. (77 persons per km^2)
Largest Cities: (1974 est.) Jakarta (5,490,000) and Surabaya (1,660,355)
Economy: Mainly agriculture, with some manufacturing
Chief Products: Agriculture—cassava, coconuts, coffee, corn, palm oil, peanuts,
  rice, rubber, spices, sugar cane, sweet potatoes, tea, tobacco; fishing—milk-
  fish, prawns, tuna; forest industry—ebony, teak; manufacturing—cement,
  glassware, petroleum products, processed foods, textiles; mining—bauxite,
  coal, natural gas, nickel, petroleum, tin
Money: Basic unit—rupiah; one hundred sen equal one rupiah
Climates: Tropical wet and dry, tropical wet

Indonesia consists of more than 13,600 islands. The climate is mostly hot
and rainy. Workers raise crops such as coffee, palm oil, rubber, sugar
cane, tea, tobacco, and rice.

Indonesians speak more than 250 Malayo-Polynesian and Papuan languages, but the Indonesian language is the official one in the country.
About 90 per cent of the people are Muslims. Many Indonesians believe
in spirits and combine ancestor and nature worship with Islam or Christianity. The famous arts of Indonesia include dances of the old royal
courts of Java and the dramatic folk dances of Bali.

Indonesia is a republic, with the central government controlling regional and local levels. The central government has encouraged foreign
investment and other means to help with severe economic problems.

Iran

Iran (69) is located in southwestern Asia. It is bordered by Russia, Afghanistan, Pakistan, Iraq, and Turkey.

Capital: Teheran (Tehran)
Official Name: Jomhori Islami Iran (Islamic Republic of Iran)
Also Called: Persia
Official Language: Farsi
Government: Islamic republic—in 1979, revolutionaries led by a Muslim religious leader overthrew the shah, or king, of Iran; a new government was set up based on the teachings of Islam with a Revolutionary Council to carry out policies; a parliament and president are elected, but have very little power
National Anthem: "Soroude Melli" ("National Anthem")
Flag: Green, white, and red horizontal stripes, with the nation's coat of arms in the center; the coat of arms is a stylized drawing of the word *Allah,* printed in Arabic
Area: 636,296 sq. mi. (1,648,000 km^2); greatest distances—northwest-southeast, 1,375 mi. (2,213 km); northeast-southwest, 850 mi. (1,370 km)
Population: 1983 estimate—40,919,000; distribution—52 per cent rural, 48 per cent urban; density—65 persons per sq. mi. (25 persons per km^2)
Largest Cities: (1976 est.) Teheran (4,716,000) and Isfahan (618,000)
Economy: Mainly agriculture and oil production
Chief Products: Agriculture—barley, corn, cotton, fruit, hides and skins, nuts, opium, tea, tobacco, tragacanth, wheat, wool; fishing—caviar; manufacturing—carpets, cement, cigarettes, silk, textiles; mining—coal, copper, iron, natural gas, petroleum, sulfur, turquoise
Money: Basic unit—rial
Climates: Subtropical moist, subtropical dry summer, steppe, desert

Iran is an ancient country sometimes called Persia. In 1979 a revolution overthrew the shah, or king, and an Islamic republic was declared.

Most Iranians live in small villages in mountain valleys. These people raise wheat, barley, and other crops wherever there is enough water to irrigate the soil. Other Iranians live in large cities, where modern buildings stand among gray mud-brick houses and ancient blue-domed mosques, or Muslim houses of worship. The rest of the people belong to wandering tribes that roam the mountains and plains with herds of livestock.

The country is one of the world's leading petroleum producers, and oil accounts for more than nine-tenths of Iran's exports. The leading craft industry is the weaving of fine Persian rugs. More than half the people of Iran cannot read and write, and most can afford little in the way of physical comforts like furniture.

77

Iraq

Iraq (70) is located in southwestern Asia. It is bordered by Turkey, Iran, Kuwait, Saudi Arabia, Jordan, and Syria.

Capital: Baghdad
Official Name: Al-Jumhuriya Al-Iraqiya (Republic of Iraq)
Official Language: Arabic
Government: Republic—president heads government and serves as commander of armed forces; president is also chairman of Revolutionary Command Council, which makes government policies; elected legislature passes laws and a Council of Ministers carries out government policies
National Anthem: "Al-Salam Al-Jumhuri" ("Salute to the Republic")
Flag: Three horizontal stripes of red, white, and black (top to bottom); a row of three green stars lies on white stripe
Area: 167,925 sq. mi. (434,924 km^2); greatest distances—north-south, 530 mi. (853 km); east-west, 495 mi. (797 km); coastline—40 mi. (64 km)
Population: 1983 estimate—14,506,000; distribution—64 per cent urban, 36 per cent rural; density—85 persons per sq. mi. (33 persons per km^2)
Largest City: (1976 est.) Baghdad (2,969,000)
Interesting Sights: Ruins of ancient cities such as Babylon and Nineveh
Economy: Mainly oil industry and agriculture
Chief Products: Agriculture—barley, cotton, dates, millet, rice, tobacco, wheat, wool; mining—petroleum; manufacturing—building materials, cotton and wool textiles, flour, leather products, soap, vegetable oil
Money: Basic unit—dinar
Climates: Subtropical dry summer, steppe, desert

Arabs make up about 80 per cent of Iraq's population, and Kurds form about 15 per cent. Roughly 95 per cent of the Iraqi people are Muslims. Most Iraqis are farmers. Their major crops include barley and dates. Iraq also has rich oil deposits. The government owns most of the large companies.

The president of Iraq is the nation's chief of state. The Revolutionary Command Council forms policy and proposes laws to the legislature. The Council of Ministers carries them out.

About a third of Iraq's people live in villages and farm their own land or lease it from the government. City people make up two-thirds of the population. Business people, craftworkers, government workers, professional people, and technicians make up the middle class. The lower class includes factory and oil workers, servants, and villagers who have moved to the cities to find better jobs.

78

Ireland

Ireland (71) is located in northwestern Europe. It is bordered by
the United Kingdom of Great Britain
and Northern Ireland.

Capital: Dublin
Official Name: Republic of Ireland
Also Called: Emerald Isle
Official Languages: English and Gaelic
Government: Republic—prime minister, appointed
   by the president, heads the government and ad-
   ministers the laws passed by the nation's two-house parliament; prime min-
   ister selects members of parliament to serve in the Cabinet and head various
   government departments
National Anthem: "The Soldier's Song"
Flag: Three vertical stripes of green, white, and orange (left to right); green rep-
   resents the country's Roman Catholics; orange, the Protestants of Ulster; and
   white, unity
Area: 27,136 sq. mi. (70,283 km^2); greatest distances—north-south, 289 mi.
   (465 km); east-west, 177 mi. (285 km); coastline—1,738 mi. (2,797 km)
Population: 1983 estimate—3,575,000; distribution—58 per cent urban, 42
   per cent rural; density—132 persons per sq. mi. (51 persons per km^2)
Largest Cities: (1979 census) Dublin (544,586) and Cork (138,267)
Interesting Sights: Celtic crosses, beautifully carved stone monuments
   throughout the country
Economy: Mainly industrial, with some agriculture
Chief Products: Agriculture—barley, dairy products, livestock, potatoes, poul-
   try, sugar beets, wheat; manufacturing—alcoholic beverages, chemicals,
   clothing, cured tobacco, machinery, metal products, paper, textiles
Money: Basic unit—Irish pound
Climate: Oceanic moist

The Emerald Isle, as Ireland is often called, is famous for its misty, green
countryside. The Irish have a reputation for being exceptionally warm-
hearted and friendly. They are also known for hospitality, close family
ties, and skill as writers and storytellers.

The nation is a republic run by a prime minister and a parliament.
Ireland is divided into 26 counties. County Kerry is famous for its mountains and the scenic Lakes of Killarney. Waterford is known for its delicate
cut glass, and Donegal for its tweed cloth.

In the 1840's, a potato famine killed hundreds of thousands of Irish.
In 1921, Ireland gained its independence from Great Britain. Since then
the Irish have built new industries, improved farming conditions, and
slowed the pace of emigration.

Israel

Israel (72) is located in southwestern Asia. It is bordered by Lebanon, Syria, Jordan, and Egypt.

Capital: Jerusalem
Official Languages: Hebrew and Arabic
Government: Republic—elected one-house parliament, the Knesset, has highest political power; prime minister, appointed by president, forms and heads the Cabinet, which runs various government departments; there is no constitution; government is based on laws passed by the parliament
National Anthem: "Hatikva" ("The Hope")
Flag: White with two blue horizontal stripes near top and bottom; blue Star of David, an ancient Jewish symbol, lies in the center of the flag
Area: 8,019 sq. mi. (20,770 km²), excluding Arab territory occupied by Israel; 34,501 sq. mi. (89,357 km²), including Arab territory; greatest distances—north-south, 256 mi. (412 km); east-west, 81 mi. (130 km); coastline—143 mi. (230 km)
Population: 1983 estimate—4,152,000; distribution—89 per cent urban, 11 per cent rural; density—518 persons per sq. mi. (200 persons per km²), excluding population of Arab territory occupied by Israel
Largest Cities: (1973 est.) Tel Aviv-Yafo (367,600), Jerusalem (326,400), and Haifa (225,800)
Interesting Sight: The Wailing Wall in East Jerusalem
Economy: Based on manufacturing and commerce, with some agriculture
Chief Products: Agriculture—citrus fruits, eggs, milk, poultry; manufacturing—chemicals, clothing and textiles, finished diamonds, machinery, metals, processed foods, transportation equipment, wood products; mining—chemical salts, copper, phosphates
Money: Basic unit—shekel
Climates: Subtropical dry summer, steppe, desert

Between 1948 and the early 1970's, over a million Jews came to Israel from other countries. Many had fled their own lands because of persecution. Arabs also claim the area that comprises Israel because they have lived there since the A.D. 600's. Jews make up about 85 per cent of the people, and Arabs about 15 per cent. Most Arabs dislike being a small minority in the Jewish country. Bitter tension and war have been a daily part of life.

The Israelis have applied great skill and energy in developing their land and industries. Israel's economy, despite poor natural resources, has grown at about 10 per cent a year.

Italy

Italy (73) is located in southern Europe. It is bordered by Switzerland, Austria, Yugoslavia, and France.

Capital: Rome
Official Name: Repubblica Italiana (Italian Republic)
Official Language: Italian
Government: Republic—has a president, a Cabinet headed by a premier, and a two-house parliament; premier determines national policy and is the most important person in the government
National Anthem: "Inno di Mameli" ("Hymn of Mameli"), by Goffredo Mameli, an Italian patriot
Flag: Three vertical stripes of green, white, and red (left to right)
Area: 116,314 sq. mi. (301,252 km^2); greatest distances—north-south, 708 mi. (1,139 km); east-west, 130 mi. (209 km); coastline—2,685 mi. (4,321 km)
Population: 1983 estimate—57,902,000; distribution—69 per cent urban, 31 per cent rural; density—497 persons per sq. mi. (192 persons per km^2)
Largest Cities: (1975 est.) Rome (2,868,248); (1972 est.) Milan (1,738,487), Naples (1,223,659), and Turin (1,172,476)
Economy: Mainly manufacturing with some commerce and agriculture
Chief Products: Agriculture—almonds, cheese, figs, fruits, grains, lemons, livestock, mulberry leaves, olives, oranges, potatoes, sugar beets, tobacco, tomatoes; manufacturing and processing—automobiles, carved ivory and marble, chemicals, clothing, leather products, machine tools, motor scooters, olive oil, petroleum products, sewing machines, ships, steel, textiles, typewriters, wine; mining—asbestos, bauxite, marble, mercury, sulfur, zinc
Money: Basic unit—lira
Climates: Oceanic moist, subtropical dry summer, highlands

Rome, the capital and largest city in Italy, has been an important center of civilization for more than 2,000 years. Almost half of Italy's universities were founded before 1350. Throughout the country are historic ruins, ancient monuments, beautiful churches and palaces, and art treasures. Opera abounds. Vatican City, the world center of the Roman Catholic Church, and a separate nation, lies within Rome.

Italy exports steel products such as automobiles and business machines. Northern Italy is one of the highly advanced industrial centers of Western Europe. In the agricultural area, the enormously rich Po Valley supplies many kinds of fruits, grains, and vegetables.

Italian politics are fraught with change. Since 1948, Italy has had more than 30 cabinet changes with few cabinets lasting more than a year.

Ivory Coast

Ivory Coast (74) is located in western Africa. It is bordered by Mali, Upper Volta, Ghana, Liberia, and Guinea.

Capital: Abidjan
Official Name: République de Côte d'Ivoire (Republic of the Ivory Coast)
Official Language: French
Government: Republic—elected president heads the government; one-house legislature is called the National Assembly; president or assembly members suggest laws, which are voted on by the legislature
National Anthem: "L'Abidjanaise" ("Hail O Land of Hope")
Flag: Vertical stripes of orange, white, and green
Area: 124,504 sq. mi. (322,463 km^2); greatest distances—north-south, 420 mi. (676 km); east-west, 411 mi. (661 km); coastline—315 mi. (507 km)
Population: 1983 estimate—9,018,000; distribution—62 per cent rural, 38 per cent urban; density—73 persons per sq. mi. (28 persons per km^2)
Largest Cities: (1978 est.) Abidjan (1,100,000) and Bouaké (230,000)
Economy: Mainly based upon agriculture, forestry, and fishing
Chief Products: Agriculture—bananas, cacao seeds, coffee, palm oil, pineapples, rubber, timber; manufacturing and processing—canned foods, electrical equipment, ships, textiles, timber products
Money: Basic unit—franc
Climates: Tropical wet and dry, tropical wet

A former French colony, Ivory Coast voted in 1958 to become a self-governing republic within the French Community. In 1960, Ivory Coast declared itself an independent republic.

Nearly all of Ivory Coast's people are black Africans. The population includes four major groups, the Akan, the Kru, the Malinke, and the Voltaic. More than 60 languages are spoken in Ivory Coast.

Most village families live in their own compounds, or groups of huts. The huts have mud walls and thatched or metal roofs. Most people in the cities also live in mud huts. A few wealthy Africans, and nearly all the non-Africans, have modern houses or apartments.

About 60 per cent of the children in Ivory Coast go to school. Most Ivorians practice ancient local religions. About 27 per cent of the people are Muslims, and about 15 per cent are Christians.

Jamaica

Jamaica (75) is an island nation in the West Indies, about 90 miles (140 kilometers) south of Cuba in the Caribbean Sea.

Capital: Kingston
Official Language: English
Government: Constitutional monarchy—prime minister leads the majority party in the country's two-house parliament and is the nation's chief executive; Cabinet members head the ministries, or executive departments, of the government; British monarch appoints the governor general who represents the monarch but has few governing powers
National Anthem: "Jamaica"
Flag: A gold diagonal cross with black triangular side panels, and green triangular panels at top and bottom
Area: 4,244 sq. mi. (10,991 km^2); greatest distances—east-west, 146 mi. (235 km); north-south, 51 mi. (82 km)
Population: 1983 estimate—2,292,000; distribution—50 per cent rural, 50 per cent urban; density—541 persons per sq. mi. (209 persons per km^2)
Largest Cities: (1978 est.) Kingston (665,050); (1970 census) Montego Bay (43,754) and Spanish Town (40,731)
Interesting Sight: Discovery Bay, site of Columbus' landing in 1494
Economy: Manufacturing, mining, agriculture, and tourism
Chief Products: Agriculture—bananas, cacao, citrus fruits, coconuts, coffee, sugar cane; manufacturing and processing—alumina, cement, chemicals, machinery, petroleum and petroleum products, sugar, textiles, tires; mining—bauxite, gypsum, silica
Money: Basic unit—Jamaican dollar
Climate: Tropical wet

The people of Jamaica include Africans, Afro-Europeans, Europeans, Syrians, and Asians (most of whom are Chinese and Indians). Jamaican business and professional people are primarily Europeans and Afro-Europeans. Many Chinese and Syrians run small shops. Large numbers of Africans and Asians work as farm laborers.

Sugar cane is the island's most important crop. Jamaicans mine bauxite, gypsum, and silica, and they manufacture many items including chemicals, cloth, fertilizer, rum, and tires. Jamaica's beautiful seaside resorts and mild climate make tourism an important industry.

During the 1960's and 1970's, Jamaica faced inflation, unemployment, poverty, and many other national problems. Many Jamaicans were dissatisfied, and their discontent sometimes led to riots and violent crime.

Japan

Japan (76) is an island nation located in the Pacific Ocean, off the east coast of Asia.

Capital: Tokyo
Official Language: Japanese
Government: Constitutional monarchy—emperor inherits the throne and is symbol of the state, but his duties are strictly ceremonial; elected members of a two-house Diet make Japan's laws; prime minister, assisted by a Cabinet, carries out government operations
National Anthem: "Kimigayo" ("The Reign of Our Emperor")
Flag: Large red sun centered on a white background
Area: 145,834 sq. mi. (377,708 km²); the four main islands—Hokkaido, Honshu, Kyushu, Shikoku—stretch about 1,200 mi. (1,900 km) from northeast to southwest; coastline—5,857 mi. (9,426 km)
Population: 1983 estimate—120,246,000; distribution—78 per cent urban, 22 per cent rural; density—824 persons per sq. mi. (318 persons per km²)
Largest Cities: (1980 census) Tokyo (8,349,209), Yokohama (2,773,822), Osaka (2,648,158), and Nagoya (2,087,884)
Interesting Sights: The Ginza, a famous shopping district in Tokyo; the Great Buddha, a huge bronze statue in Kamakura; the National Stadium of Tokyo
Economy: Mainly based upon manufacturing and trade
Chief Products: Agriculture—rice, wheat, barley, potatoes, soybeans, sweet potatoes, fruits, tea, tobacco; fishing industry—sardines, cod, mackerel, squid, tuna, shellfish, halibut, salmon, whales; manufacturing—transportation equipment, machinery, electronic equipment, precision instruments, iron and steel, chemicals, textiles; mining—coal, zinc, manganese, copper, lead, silver
Money: Basic unit—yen
Climates: Continental moist, subtropical moist

Mountains and hills cover about six-sevenths of Japan and make it a very beautiful country. It is poor, however, in land and minerals. The skillful, hard-working people are their country's greatest resource. They have used their land and minerals wisely and have imported what they needed to make Japan one of the world's great industrial nations.

Today, Japan is a democracy. It also is a land of contrast—of old and new, of East and West. Graceful temples hundreds of years old stand near modern steel and concrete buildings. The traditional Japanese *no* and *kabuki* dramas may still be seen, along with the latest motion pictures. Sports fans enjoy sumo, an ancient Japanese style of wrestling, as well as baseball, Japan's most popular sport. Women still wear traditional kimonos, plus the latest Western styles.

Jordan

Jordan (77) is located in south-western Asia. It is bordered by Syria, Iraq, Saudi Arabia, and Israel.

Capital: Amman
Official Name: Al-Mamlakah Al-Udri-niyah Al-Hashimiyah (Hashemite King-dom of Jordan)
Official Language: Arabic
Government: Constitutional monarchy—king se-lects a prime minister to govern the nation and chooses members of the Cabinet; the king also appoints members of the National Assembly, Jordan's two-house legislature
National Anthem: "Al-Salam Al-Malaki" ("The Royal Salute")
Flag: Horizontal black, white, and green stripes; a seven-pointed star on a red triangle represents basic Islamic beliefs; the flag's four colors stand for four periods in Arab history
Area: 37,738 sq. mi. (97,740 km^2); greatest distance—north-south and east-west, about 230 mi. (370 km); coastline—4 mi. (6.4 km)
Population: 1983 estimate—2,460,000; distribution—58 per cent rural, 42 per cent urban; density—65 persons per sq. mi. (25 persons per km^2)
Largest Cities: (1979 census) Amman (648,587) and Az Zarqa (215,687)
Interesting Sights: Historical ruins such as the Greek and Roman city of Jar-ash, the Nabataean city of Petra carved in a rock cliff, and the crusader castles at Al Karak and Ajlun; Biblical sites such as Jerusalem, Bethlehem, Hebron, and Jericho
Economy: Mainly agricultural, with some developing industry
Chief Products: Barley, fruit, goats, lentils, sheep, vegetables, wheat
Money: Basic unit—dinar
Climates: Steppe, desert

Jordan is an Arab kingdom directly east of Israel. The River Jordan di-vides the country into two parts: the east and west banks. Since the Arab-Israeli war of 1967, Israel has controlled the west bank, where about a fourth of Jordan's population lives. King Hussein of Jordan gave up his claim to the west bank in 1974, but the problem remains of what to do with the Arab refugees still living there and with those who came to the east bank following the war. The increased number of refugees on the east bank has weakened Jordan's economy.

About 60 per cent of Jordanians 15 years old and older can read and write. Jordan has three main groups of people. The villagers who farm for a living make up the largest group. The next consists of city people, who work in trade and manufacturing. The Bedouins, the smallest group, wander in the desert tending herds of livestock.

85

Kenya

Kenya (78) is located in eastern Africa. It is bordered by Sudan, Ethiopia, Somalia, Tanzania, and Uganda.

Capital: Nairobi
Official Name: Jamhuri ya Kenya (Republic of Kenya)
Official Language: Swahili
Government: Republic—headed by elected president who is assisted by about 20 appointed Cabinet ministers; one-house legislature, the National Assembly, makes the country's laws
National Anthem: "Wimbo wa Taifa" ("Anthem of the Nation")
Flag: Three horizontal stripes of black, red, and green (top to bottom); two white stripes separate the red from the black and the green; a shield and spears, representing the defense of freedom, lies in the center of the flag; the black stripe stands for the Kenyan people, the red for their struggle for independence, and the green for agriculture
Area: 224,961 sq. mi. (582,646 km^2); greatest distances—north-south, 640 mi. (1,030 km); east-west, 560 mi. (901 km); coastline—284 mi. (457 km)
Population: 1983 estimate—18,185,000; distribution—86 per cent rural, 14 per cent urban; density—88 persons per sq. mi. (34 persons per km^2)
Largest Cities: (1979 census) Nairobi (835,000) and Mombasa (342,000)
Interesting Sights: Fascinating wildlife, including elephants, giraffes, rhinoceroses, and zebras
Economy: Mainly agricultural, with growing manufacturing
Chief Products: Agriculture—coffee, corn, tea, sisal, wheat, sugar cane, meat; manufacturing—cement, chemicals, light machinery, textiles, processed foods, petroleum products
Money: Basic unit—Kenya shilling
Climates: Steppe, desert, tropical wet and dry

Almost all of Kenya's people are black Africans. Corn is the basic food, and beer is a popular beverage. Dancing is a favorite form of recreation throughout the country, and Kenyans have created highly artistic dances that are performed during ceremonies such as birth celebrations, marriages, and funerals.

Formerly ruled by the British, Kenya gained its independence in 1963 after a long fight. The government then took over many farms and businesses owned by non-Africans and sold or rented them to Africans. The economy operates as a free enterprise system.

Following independence, the public school system was quickly expanded. The government moved to overcome divisions among ethnic groups, giving the people a sense of national unity.

Kiribati

Kiribati (79) is an island nation in the southwest Pacific Ocean.

Capital: Tarawa
Official Languages: English and
  Gilbertese
Government: Republic—headed by
  elected president; one-house leg-
  islature, the House of Assembly,
  makes nation's laws
National Anthem: "Stand, Kiribati"
Flag: Alternating blue and white horizontal stripes fill bottom third of the flag;
  top two-thirds shows a golden eagle flying above a red and gold flaming
  half-sun, all on a red field
Area: 278 sq. mi. (719 km^2)
Population: 1981 estimate—61,000; density—219 persons per sq. mi. (85 per-
  sons per km^2)
Largest City: (1978 census) Tarawa (17,921)
Economy: Mainly agricultural, with some trade and limited mining
Chief Products: Agriculture—copra or dried coconut meat; mining—phosphate
Money: Basic unit—Australian dollar
Climate: Tropical wet

Thirty-three islands make up Kiribati—Ocean Island, the 16 Gilbert Islands, the 8 Phoenix Islands, and 8 of the Line Islands. The islands are
scattered over about 2 million square miles (5 million square kilometers)
of ocean. Yet the total land area of Kiribati is only 278 square miles (719
square kilometers).

The people of Kiribati are Micronesians who call themselves I-Kiribati.
Their ancestors have lived on the islands since before A.D. 1400. The I-
Kiribati enjoy a beautiful tropical climate, with day and night tempera-
tures of about 80°F (27°C) the year round.

Many of the islands of Kiribati are atolls, or ring-shaped reefs that en-
close a lagoon. Villages dot the islands. They consist of 10 to 170 houses
clustered around a church or meeting hall. Many of the houses are made
of wood and leaves from palm trees.

The I-Kiribati raise most of their own food, including bananas, coconut,
breadfruit, papaya, sweet potatoes, and taro. The islanders also raise pigs
and chickens and catch fish for food.

Korea, North

North Korea (80) is located in eastern Asia. It is bordered by China, Russia, and South Korea.

Capital: Pyongyang
Official Name: Choson-minjujuui-inmin-konghwaguk (Democratic People's Republic of Korea)
Official Language: Korean
Government: Communist dictatorship—country's Communist Party, the Korean Workers' Party, holds the real political power; the party makes all laws and chooses all candidates for election; nation's president heads the Central People's Committee, the country's most powerful government body
National Anthem: No official anthem; "Aegug-ka" ("National Anthem") used as unofficial national anthem
Flag: A blue horizontal stripe across top and bottom; red in the center separated from blue by two narrow white bands; a red star, representing Communism, lies in a white circle on red portion of flag
Area: 46,540 sq. mi. (120,538 km^2), including islands and excluding the 487-sq.-mi. (1,262-km^2) demilitarized zone; greatest distances—north-south, 370 mi. (595 km); east-west, 320 mi. (515 km); coastline—665 mi. (1,070 km)
Population: 1983 estimate—19,291,000; distribution—60 per cent urban, 40 per cent rural; density, 414 persons per sq. mi. (160 persons per km^2)
Largest Cities: (1971 est.) Pyongyang (2,500,000) and Hamhung (525,000)
Economy: Based upon manufacturing and mining, with limited agriculture
Chief Products: Agriculture—barley, corn, millet, rice, wheat; manufacturing—chemicals, iron and steel, machinery, textiles; mining—graphite, magnesium, tungsten; fishing—shellfish and many kinds of saltwater fish
Money: Basic unit—won
Climate: Continental moist

North Korea lies on the Korean Peninsula, with South Korea. Formerly one country, North and South Korea divided following a struggle between Communist and non-Communist nations. North Korea's government is a Communist state called the Democratic People's Republic.

North Korea's economy used to be based on agriculture, but since the 1950's the country has made special efforts to develop industries. Now it has one of the world's fastest growing economies.

Koreans, with their broad faces, straight black hair, olive-brown skin, and dark eyes, resemble the Chinese and Japanese. Many North Koreans still wear traditional dress, but most have plain, uniformlike clothing. Their homes usually have little furniture. Rice is the basic food, and meat and dairy products are scarce in North Korea. See also **Korea, South**.

Korea, South

South Korea (81) is located in north-eastern Asia. It is bordered by North Korea.

Capital: Seoul
Official Name: Taehan-minguk (Republic of Korea)
Official Language: Korean
Government: Republic—president heads the government and appoints a prime minister to carry out government operations; members of the National Assembly, the country's legislature, are elected by the people
National Anthem: "Aegug-ka" ("National Anthem")
Flag: White background with a circle of red and blue symbolizing the balance in the universe between opposites—such as night and day, and life and death; four black trigrams called *kwae* are around the circle and represent the four seasons
Area: 38,025 sq. mi. (98,484 km²), including islands and excluding the 487-sq.-mi. (1,262-km²) demilitarized zone; greatest distances—north-south, 300 mi. (480 km); east-west, 185 mi. (298 km); coastline—819 mi. (1,318 km)
Population: 1983 estimate—39,275,000; distribution—55 per cent urban, 45 per cent rural; density—1,033 persons per sq. mi. (399 persons per km²)
Largest Cities: (1975 census) Seoul (6,879,464), Pusan (2,450,125), Taegu (1,310,768), and Inchon (800,007)
Economy: Mainly manufacturing and agriculture
Chief Products: Agriculture—barley, beans, potatoes, rice, wheat; manufacturing—chemicals, machinery, processed foods, textiles; mining—anthracite, fluorite, graphite, iron ore, salt, tungsten; fishing—shellfish and many kinds of saltwater fish, including anchovy and herring
Money: Basic unit—won
Climates: Continental moist, subtropical moist

South Korea shares the Korean Peninsula with North Korea. The nation is a republic headed by a president, a state council, and a legislature. At least 30,000 years ago, ancestors of the Korean people lived in what is now Korea. Since ancient times, other countries like China and Japan have often had strong influence on the peninsula.

Since the late 1940's, when Korea was divided, more than 24,000 manufacturing firms have come into being in South Korea. The chief products are chemicals, machinery, processed foods, and textiles. South Korea enjoys one of the world's fastest growing economies.

Most South Koreans who follow a religion are Buddhists. But Confucianism, which is more a philosophy than a religion, traditionally has comprised the principal set of beliefs. See also **Korea, North**.

Kuwait

Kuwait (82) is located in south-
western Asia. It is bordered by
Iraq, Iran, and Saudi Arabia.

Capital: Kuwait
Official Name: Dowlat al Kuwait
  (State of Kuwait)
Official Language: Arabic
Government: Emirate—emir is the head
  of state and appoints the prime minister; prime
  minister chooses ministers who are confirmed
  by the emir; 50-member legislature, the National Assembly, makes the country's laws and is elected by the country's adult males; police officers, military
  personnel, and women are not eligible to vote
Flag: Horizontal green, white, and red stripes (top to bottom) join a black, trap-
  ezoid-shaped stripe at the flagstaff
Area: 7,780 sq. mi. (20,150 km²), including offshore islands; greatest distances—
  east-west, 95 mi. (153 km); north-south, 90 mi. (145 km); coastline—120
  mi. (193 km)
Population: 1983 estimate—1,610,000; distribution—88 per cent urban, 12 per
  cent rural; density—207 persons per sq. mi. (80 persons per km²)
Largest Cities: (1975 census) Hawalli (130,565), As Salimiyah (113,943), and
  Kuwait (78,116)
Economy: Based almost entirely upon oil
Chief Products: Petroleum, natural gas
Money: Basic unit—dinar
Climate: desert

A poor, backward country until 1946, Kuwait is today one of the rich
and progressive countries in the world. With the wealth gained by selling
oil, Kuwait's rulers have turned a barren desert country into a prosperous
welfare state. It has free primary and secondary education, free health
and social services, and no income tax. The nation gives financial aid to
Arab and non-Arab countries in Africa and Asia through the Kuwait Fund
for Arab Economic Development.

Kuwait is governed by a ruler called an emir, who appoints a prime
minister. The prime minister chooses the ministers, whom the emir con-
firms.

Most people of Kuwait are Arabs and Muslims. Arabic is the official
language, and although Islam is the declared state religion, religious dis-
crimination is forbidden.

Laos

Laos (83) is located in Southeast Asia. It is bordered by China, Vietnam, Cambodia, Thailand, and Burma.

Capital: Vientiane
Official Language: Lao
Government: Socialist republic (Communist dictatorship)—called Provisional Government of the National Union; a 42-member National Political Council advises the government
National Anthem: "Pheng Sat" ("National Music")
Flag: A red horizontal stripe at the top and the bottom, and a blue horizontal stripe in the center; a white circle appears in the center of the flag; the red symbolizes the blood and soul of the Laotian people; the blue stands for prosperity; the white circle represents the promise of a bright future
Area: 91,429 sq. mi. (236,800 km^2); greatest distances—northwest-southeast, 650 mi. (1,046 km); northeast-southwest, 315 mi. (510 km)
Population: 1983 estimate—3,995,000; distribution—85 per cent rural, 15 per cent urban; density—44 persons per sq. mi. (17 persons per km^2)
Largest Cities: (1973 est.) Vientiane (174,229), Savannakhet (50,691), and Pakxé (44,860)
Economy: Mainly agriculture
Chief Products: Benzoin, cardamom, cattle, cinchona, citrus fruits, coffee, corn, cotton, leather goods, opium, pottery, rice, silk, silverwork, tea, teak, tin, tobacco
Money: Basic unit—kip
Climates: Subtropical moist, tropical wet and dry, tropical wet

In 1954, an international agreement recognized the French colony of Laos as an independent, neutral nation. Civil war broke out in 1960 between Laotian government troops and the Communist-led Pathet Lao forces. In 1975, the Pathet Lao won the war and took control of the country.

Laos has practically no manufacturing industries. Almost all its people are farmers. Most of them raise rice along the Mekong River and its tributaries, and corn, cotton, rice, and tobacco in the highlands. Most Laotians live in houses perched on wooden posts 6 to 8 feet (1.8 to 2.4 meters) aboveground. The houses have covered porches, bamboo floors and walls, and thatched roofs.

Most village dwellers are poor. The songs of traveling ballad singers are often the only source of news. Only about a tenth of Laotians can read and write. Most are Buddhists, and much of the country's social life centers around Buddhist festivals and holidays.

Lebanon

Lebanon (84) is located in south-western Asia at the eastern end of the Mediterranean Sea. It is bordered by Syria and Israel.

Capital: Beirut

Official Language: Arabic

Government: Republic—headed by president who appoints prime minister; prime minister has less power than president, but traditionally both have worked as a team; Council of Ministers, chosen by prime minister with president's approval, carries out government operations; one-house parliament, the National Assembly, makes the country's laws

National Anthem: "Kulluna lil watan lil ula lil alam" ("All of Us for the Country, Glory, and Flag")

Flag: Three horizontal stripes—red, white, and red; a cedar tree on the white stripe symbolizes holiness, eternity, and peace

Area: 4,015 sq. mi. (10,400 km^2); greatest distances—north-south, 120 mi. (193 km); east-west, 50 mi. (80 km); coastline—130 mi. (209 km)

Population: 1983 estimate—3,404,000; distribution—76 per cent urban, 24 per cent rural; density—847 persons per sq. mi. (327 persons per km^2)

Largest City: (1974 est.) Beirut (702,000)

Economy: Trade and commerce, with some manufacturing and agriculture

Chief Products: Agriculture—apples, cherries, cucumbers, grapes, lemons, oranges, peaches, tomatoes; manufacturing—cement, chemicals, electric appliances, furniture, processed foods, textiles

Money: Basic unit—pound

Climate: Subtropical dry summer

Lebanon has been a world transportation and trade center for about 4,000 years. The cedars of Lebanon have grown on its mountain slopes since Biblical times. Few, however, remain today. There are stony paths that Jesus once used. The ruins of Phoenician ports, Roman temples, and castles built by the crusaders still stand there.

Lebanon is considered an Arab country. About half of the people are Muslims, but most of the rest are Christians. The religious differences are reflected in government posts. Bitter tension between the two groups caused a bloody civil war in the mid-1970's.

Trade and service industries are Lebanon's major sources of income. It exports farm products and textile fibers. Oil refining is also a major industry. Lebanon is also a large financial center, with about 100 banks. Lebanon's scenic beauty, its mild climate, the ruins of the past, the hospitality of its villages—all lure many tourists.

Lesotho

Lesotho (85) is surrounded by the Republic of South Africa.

Capital: Maseru
Official Name: Kingdom of Lesotho
Formerly Called: Basutoland
Official Languages: English and Sesotho
Government: Constitutional monarchy—prime minister and cabinet direct the government; prime minister is leader of the majority party in the 93-member legislature, called the National Assembly; 71 members of the assembly are appointed by the king on advice of prime minister; 22 members are local chiefs; king, who is also Paramount Chief of Lesotho, has some powers in tribal matters
National Anthem: "Lesotho Fatse La Bo-Ntata Rona" ("Lesotho Our Fatherland")
Flag: Narrow vertical stripes of green and red near the hoist; a white straw hat lies on a blue field in the center
Area: 11,720 sq. mi. (30,355 km^2); greatest distances—north-south, 140 mi. (224 km); east-west, 140 mi. (224 km)
Population: 1983 estimate—1,442,000; distribution—95 per cent rural, 5 per cent urban; density—124 persons per sq. mi. (48 persons per km^2)
Largest City: (1976 census) Maseru (38,440)
Economy: Mainly agricultural
Chief Products: Agriculture—beans, corn, hides and skins, livestock, millet, mohair, oats, peas, wheat, wool
Money: Basic unit—South African rand
Climates: Subtropical moist, steppe

Lesotho is called the Switzerland of Southern Africa because of its beautiful mountain scenery. But it is a poor country. It has no manufacturing industries and almost no minerals. Many of its people go to South Africa to find jobs or to fulfill 6- or 9-month work contracts.

Lesotho is the former British colony of Basutoland. It became independent in 1966. Most of Lesotho's people are black Africans called Basotho or Basuto. They are strong, independent people who raise livestock and food crops. The wealth of a family is usually measured by the number of cattle it owns, and cattle are often used instead of money.

In the villages, family groups build their thatched-roof huts around a cattle pen. The women hoe and weed the land, harvest the crops, and build the houses. The men plow the land and look after the sheep, cattle, and goats. Missionaries run most of the schools, and about 40 per cent of the people can read and write.

Liberia

Liberia (86) is located on the west coast of Africa. It is bordered by Guinea, Ivory Coast, and Sierra Leone.

Capital: Monrovia
Official Name: Republic of Liberia
Official Language: English
Government: Military rule—in 1980 a military revolt was staged; the Constitution and legislature were abolished and all elections suspended; People's Redemption Council was established to run the government; leader of revolt became chairman of the council and head of state
Flag: Six red and five white horizontal stripes that represent the 11 signers of the Liberian Declaration of Independence; a white star appears on a dark blue canton in the upper left corner
Area: 43,000 sq. mi. (111,369 km²); greatest distances—east-west, 230 mi. (370 km); north-south, 210 mi. (338 km); coastline—315 mi. (507 km)
Population: 1983 estimate—2,071,000; distribution—67 per cent rural, 33 per cent urban; density—49 persons per sq. mi. (19 persons per km²)
Largest Cities: (1974 census) Monrovia (204,210) and Buchanan (23,994)
Economy: Mainly mining, with substantial agriculture and trade
Chief Products: Agriculture—bananas, cassava, coffee, rice, rubber; forest products—kola nuts, palm oil, piassava; mining—diamonds, gold, iron ore
Money: Basic unit—United States dollar
Climates: Tropical wet and dry, tropical wet

Liberia is the oldest black republic in Africa. It was founded in 1822 as a home for freed black slaves from the United States. Today, only about 5 per cent of all Liberians are Americo-Liberians, or descendants of the original American settlers. Most of the rest are blacks who belong to about 20 different ethnic groups. Most Americo-Liberians speak English, are Protestant, and live in coastal cities. Most of the other people speak various African languages and make a living as farmers or as workers on large rubber plantations.

Many Liberian city dwellers dress much like Americans and live in houses made of brick, wood, or concrete. Tennis, soccer, and movies are popular forms of recreation. The people in rural areas live in round huts with mud walls and thatched roofs. Farm families work together to raise their crops, which include cassava, rice, taro, and yams. Exports of iron ore and rubber provide Liberia with most of its income. The country also earns money from registration fees paid by foreign shipowners who register their ships in Liberia in order to take advantage of lower taxes.

Libya

Libya (87) is located on the northern coast of Africa. It is bordered by Egypt, Sudan, Chad, Niger, Algeria, and Tunisia.

Capital: Tripoli
Official Name: People's Socialist Libyan Arab Jamahiriya
Official Language: Arabic
Government: Jamahiriya (the Libyan government's name for a kind of republic)—a military revolt in 1969 overthrew Libya's monarchy; the leader of the revolution is presently the country's most powerful leader; a General Secretariat and a General People's Congress assist him in carrying out government operations
Flag: Entirely green—the traditional color of Islam, the religion of most Libyans
Area: 679,362 sq. mi. (1,759,540 km^2); greatest distances—north-south, 930 mi. (1,497 km); east-west, 1,050 mi. (1,690 km); coastline—1,047 mi. (1,685 km)
Population: 1983 estimate—3,226,000; distribution—70 per cent rural, 30 per cent urban; density—5 persons per sq. mi. (2 persons per km^2)
Largest Cities: (1973 census) Tripoli (735,083) and Benghazi (337,423)
Economy: Based mainly on the production of oil
Chief Products: Agriculture—barley, citrus fruits, dates, livestock, olives, wheat; mining—oil
Money: Basic unit—dinar
Climates: Steppe, desert

Libya is a country covered largely by desert. The Sahara extends across more than 90 per cent of its total land area. But large oil deposits have brought Libya considerable wealth. Oil revenues have enabled the country to build electric power plants, irrigation systems, and other development projects. Oil provides most government income, but the oil industry employs only a small percentage of Libya's workers. About 80 per cent of the people are farmers.

Most Libyans live in a narrow strip of fertile land along the Mediterranean coast. Some live in desert oases. Others are nomads, who roam from place to place with herds of livestock. Livestock, including cattle, sheep, and goats, are Libya's chief agricultural products. Major crops include barley, citrus fruits, dates, olives, and wheat.

Nearly all Libyans are Arabs or are of mixed Arab and Berber ancestry. Most of the people speak Arabic and are Muslims. Some Libyans wear Western-style clothing, but most wear traditional Arab robes. About half the school-age children in Libya attend school, and about a third of the population can read and write.

Liechtenstein

Liechtenstein (88) is located in south-central Europe. It is bordered by Austria and Switzerland.

Capital: Vaduz
Official Name: Fürstentum Liechtenstein (Principality of Liechtenstein)
Official Language: German
Government: Principality—ruled by a
prince who appoints a prime minister to direct
the government; 15-member parliament, called
the Landtag, is elected by the nation's men; it passes laws, prepares the
national budget, sets tax rates, and appoints four government councilors to
assist the prime minister
Flag: Two stripes, the upper one blue (for the sky), the lower one red (for the
glow of evening fires); a gold crown (for the unity of the royalty and the
people) appears in the upper left-hand corner
Area: 61 sq. mi. (157 km^2); greatest distances—north-south, 17.4 mi. (28 km);
east-west, 7 mi. (11 km)
Population: 1983 estimate—28,000; distribution—72 per cent rural, 28 per
cent urban; density—461 persons per sq. mi. (178 persons per km^2)
Largest City: (1977 est.) Vaduz (4,704)
Economy: Based mainly on industry, trade, and agriculture
Chief Products: Agriculture—beef and dairy cattle; manufacturing—handicrafts,
precision instruments, cotton textiles
Money: Basic unit—Swiss franc
Climate: Highlands

Liechtenstein is a tiny nation with close ties to neighboring Switzerland. Its people speak German in a Swiss dialect and use Swiss money. Switzerland operates Liechtenstein's postal, telephone, and telegraph systems.

Neat wooden houses are characteristic of the small villages where most Liechtensteiners live. Many of the people wear traditional embroidered clothing. More than half of all Liechtensteiners work as craftspeople or in small factories. They make leather goods, pottery, cotton textiles, precision instruments, and other products. About a fifth of the people are farmers. They grow corn, potatoes, and other vegetables in the flat lowlands next to the Rhine River. Grapes and fruits are grown in the upland areas, and beef and dairy cattle graze in Alpine meadows.

More than 5,000 foreign businesses have headquarters in Liechtenstein because of the country's low tax rates. In addition to tax revenue, the government earns money by selling postage stamps. Stamp collectors throughout the world prize Liechtenstein's beautiful stamps.

Luxembourg

Luxembourg (89) is located in north-western Europe. It is bordered by Belgium, West Germany, and France.

Capital: Luxembourg
Official Name: In French, Grand-Duché de Luxembourg; in German, Grossherzogtum Luxemburg (Grand Duchy of Luxembourg)
Official Languages: French and German (common language—Letzburgesch)
Government: Constitutional monarchy—the grand duke (or duchess) of the House of Nassau is chief executive and appoints a Cabinet consisting of a prime minister and seven other ministers, each in charge of one or more government departments; 21 members of the Council of State are appointed for life by grand duke or duchess; Chamber of Deputies passes all laws
Flag: Three horizontal stripes of red, white, and blue (top to bottom); the colors come from the coat of arms of Luxembourg
Area: 998 sq. mi. (2,586 km^2); greatest distances—north-south, 55 mi. (89 km); east-west, 35 mi. (56 km)
Population: 1983 estimate—368,000; distribution—68 per cent urban, 32 per cent rural; density—368 persons per sq. mi. (142 persons per km^2)
Largest City: (1979 est.) Luxembourg (79,600)
Economy: Mainly mining and industry
Chief Products: Agriculture—cattle, grapes, oats, potatoes, wheat; mining—iron ore; manufacturing—ceramics, iron, machinery, paints, steel, wine
Money: Basic unit—franc
Climate: Oceanic moist

Luxembourg is a small nation that ranks as one of the highly industrialized countries in the world. Rich deposits of iron ore are its most valuable resource. Seven steel mills in southern Luxembourg employ about a third of the country's workers.

In spite of its industrial character, Luxembourg is also a land of scenic beauty and charm. Rolling hills and dense forests cover much of the country. Whitewashed houses cluster around ancient castles and churches in the small towns and villages where about a third of the people live. Farmers grow grapes on terraced land along the Moselle River, and fields of oats, potatoes, and wheat mark the countryside.

Most Luxembourgers speak Letzburgesch, a form of German. About 95 per cent are Roman Catholics. Almost all adult Luxembourgers can read and write. In general, Luxembourgers have more modern conveniences and better food and housing than many other Europeans. Ham and freshwater fish, especially trout, are favorite foods.

Madagascar

Madagascar (90) is an island nation located in the Indian Ocean about 240 miles (386 kilometers) off the east coast of Africa.

Capital: Antananarivo
Official Name: Democratic Republic of Madagascar
Formerly Called: Malagasy Republic
Official Languages: Malagasy and French
Government: Republic (military rule)—government body called a military directory makes the country's laws and runs the government; leader of the directory serves as chief of government and chief of state; the nation does have an elected parliament
Flag: A white vertical stripe appears at the left, with a red horizontal stripe over a green one at the right; white is for purity, red for sovereignty, and green for hope
Area: 226,658 sq. mi. (587,041 km²); greatest distances—north-south, 980 mi. (1,580 km); east-west, 360 mi. (579 km); coastline—2,600 mi. (4,180 km)
Population: 1983 estimate—9,387,000; distribution—82 per cent rural, 18 per cent urban; density—41 persons per sq. mi. (16 persons per km²)
Largest Cities: (1977 est.) Antananarivo (484,000) and Fianarantsoa (73,000)
Economy: Mainly agricultural, with some mining and very limited industry
Chief Products: Agriculture—cattle, coffee, rice, vanilla; mining—graphite, mica, semiprecious stones; manufacturing—chemicals, cigarettes, sisal, sugar, textiles
Money: Basic unit—franc
Climates: Steppe, tropical wet and dry, tropical wet

Madagascar is populated by two major groups: people of black African descent and people of Indonesian descent. Most of the blacks live along the coast and most of the people of Indonesian background live in the central highlands. People throughout Madagascar speak Malagasy, a language that resembles Malay and Indonesian. Malagasy and French are the country's two official languages.

About four-fifths of all Madagascar's workers are farmers or herders. Madagascar is the world's leading producer of vanilla, but coffee ranks as its most valuable export. Most of the farmers raise cattle, and for food many grow rice, corn, cassava, and potatoes. The people also eat fruit and sometimes meat or fish.

About a third of the people of Madagascar are Christians. Most of the rest practice tribal religions that center around worship of spirits and ancestors. Nearly half the school-age children in Madagascar attend elementary schools, but only about 4 per cent attend secondary schools.

Malawi

Malawi (91) is located in south-eastern Africa. It is bordered by Tanzania, Mozambique, and Zambia.

Capital: Lilongwe
Formerly Called: Nyasaland
Official Languages: Chichewa and English
Government: Republic—elected president serves as head of state and chief executive; president appoints a Cabinet to help govern the country; in 1970, a constitutional amendment made Hastings Kamuzu Banda president for life; people elect 60 members of the country's parliament and the president appoints the remaining 18 members
Flag: Three horizontal stripes of black, red, and green (top to bottom); a red rising sun lies on the black stripe
Area: 45,747 sq. mi. (118,484 km^2); greatest distances—north-south, 520 mi. (837 km); east-west, 100 mi. (160 km)
Population: 1983 estimate—6,427,000; distribution—66 per cent rural, 34 per cent urban; density—140 persons per sq. mi. (54 persons per km^2)
Largest Cities: (1977 census) Blantyre (229,000), Lilongwe (102,924), and Zomba (16,000)
Economy: Almost entirely agricultural
Chief Products: Agriculture—coffee, cotton, hides and skins, peanuts, tea, tobacco, tung oil; manufacturing and processing—bricks, cement, cotton goods, furniture, soap
Money: Basic unit—kwacha
Climate: Tropical wet and dry

Malawi is a small scenic country of lakes, mountains, forests, and grasslands. It lies on the western shore of Lake Nyasa. Most of Malawi's people are black Africans who belong to various ethnic groups. Europeans and Asians make up small minorities. English and an African language called Chichewa are Malawi's official languages. Other African languages are spoken as well. About two-thirds of Malawi's people live in rural areas. Round or oblong houses with mud walls and thatched roofs are typical.

Malawi has few manufacturing industries and no important mineral deposits. Only about a third of the country's land is suitable for farming. Thousands of Malawi men work in mines in nearby Zambia, Zimbabwe, and South Africa. While they are away from home, their families grow food crops such as corn, sorghum, and millet. Tea, the leading export crop, is grown on estates owned by Europeans.

Malaysia

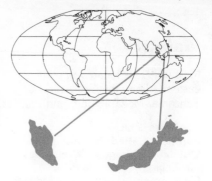

Malaysia (92) is located in Southeast Asia. It covers the southern portion of the Malay Peninsula and most of the northern portion of the island of Borneo. It is bordered by Thailand and Indonesia.

Capital: Kuala Lumpur
Official Language: Bahasa Malaysia
Government: Constitutional monarchy—king serves as head of state, but his duties are largely ceremonial; prime minister is the top government official and selects a Cabinet to assist in carrying out government operations; an elected two-house Parliament makes country's laws
National Anthem: "Negara Ku" ("My Country, My Native Land")
Flag: A yellow crescent and star lie on a blue background in the upper left corner; the crescent represents Islam; the star's 14 points and the flag's 14 red and white stripes symbolize Malaysia's 14 original states
Area: 127,317 sq. mi. (329,749 km^2); greatest distances—east-west, 1,350 mi. (2,160 km); north-south, 450 mi. (720 km); coastline—2,132 mi. (3,431 km)
Population: 1983 estimate—14,554,000; distribution—71 per cent rural, 29 per cent urban; density—114 persons per sq. mi. (44 per km^2)
Largest Cities: (1970 census) Kuala Lumpur (451,728) and George Town (270,019)
Economy: Mainly agricultural, with some manufacturing and mining
Chief Products: Agriculture—rubber, rice, cacao seeds, coconuts, palm oil, pepper, pineapples, timber; manufacturing—cement, chemicals, textiles, rubber goods, processed foods; mining—tin, petroleum, bauxite, copper, gold
Money: Basic unit—ringgit (sometimes called Malaysian dollar)
Climate: Tropical wet

Malaysia is a tropical land that ranks as the world's leading producer of natural rubber and tin. Factories that process these materials are among Malaysia's chief industries. But less than 10 per cent of the country's workers are industrial laborers. About half the people depend on agriculture to earn a living.

Malays and Chinese are the two largest population groups in Malaysia. Malays dominate the country's government, but the Chinese control the economy. Most of the Malays are farmers who grow rice and other crops for their own use. They practice Islam, Malaysia's official religion. Most of the Chinese live in cities, and many are merchants or bankers. Malaysia's chief minorities are Indians and people called Dyaks and Kadazans. Most of the Indians are Hindus. Many are plantation workers. The Dyaks and Kadazans practice local traditional religions.

Maldives

Maldives (93) is a small nation composed of about 2,000 islands located in the Indian Ocean.

Capital: Male
Official Name: Divehi Raajje (Republic of Maldives)
Official Language: Divehi
Government: Republic—president is elected by the Majlis, the country's one-house legislature; president heads the government assisted by a nine-member Cabinet; people elect 46 members of the Majlis, and the president appoints the remaining eight members
Flag: A white crescent on a dark green rectangle with a red border; the colors and the crescent on the flag stand for Islam
Total Land Area: 115 sq. mi. (298 km^2); greatest distances—north-south, 550 mi. (885 km); east-west, 100 mi. (161 km)
Population: 1983 estimate—159,000; distribution—86 per cent rural, 14 per cent urban; density—1,383 persons per sq. mi. (534 persons per km^2)
Largest City: (1978 census) Male (29,555)
Economy: Based on the government-controlled fishing industry
Chief Products: Agriculture—breadfruit, coconuts, papaya, pineapples, pomegranates, yams; fishing—bonito, tuna; handicrafts—coir yarn, cowrie shells, lacquerware, woven mats
Money: Basic unit—rupee
Climate: Tropical wet and dry

Maldives is a nation of green tropical islands, white sand beaches, and clear lagoons. Most of its inhabitants are descendants of Sinhalese people who originally came from Sri Lanka. They speak a language called Divehi, and nearly all of them are Sunni Muslims.

Life in the Maldives has changed little in hundreds of years. To catch tuna and bonito, the men go out to sea every day in long, narrow boats made of coconut logs or other timber. Most of the fish are dried and exported. But for their own food, the people cook and smoke some fish over fires. Maldivians grow most of their other foods, including coconuts, papayas, pineapples, pomegranates, and yams. Maldivian women weave reed mats and make yarn and rope out of coconut husk fibers called coir. The men make lacquerware. The people barter for things they need other than what they grow or make themselves. They use sailboats as their chief means of transportation. In addition to dried fish, the Maldives exports coir yarn, fish meal, dried coconut meat, and cowrie shells.

Mali

Mali (94) is located in western Africa. It is bordered by Algeria, Niger, Upper Volta, Ivory Coast, Guinea, Senegal, and Mauritania.

Capital: Bamako
Official Name: République du Mali (Republic of Mali)
Formerly Called: French Sudan
Official Language: French
Government: Republic (military rule)—military officers overthrew the government in 1968, dissolved all political organizations, and jailed the president and other government officials; Military Committee of National Liberation governs the country; committee's president serves as Mali's head of state and head of government; Council of Ministers carries out the committee's policies
Flag: Three vertical stripes of green, gold, and red (left to right) which symbolize devotion to republicanism and the Declaration of the Rights of Man
Area: 478,767 sq. mi. (1,240,000 km^2); greatest distances—east-west, 1,150 mi. (1,851 km); north-south, 1,000 mi. (1,609 km)
Population: 1983 estimate—7,481,000; distribution—83 per cent rural, 17 per cent urban; density—16 persons per sq. mi. (6 per km^2)
Largest Cities: (1976 census) Bamako (404,022); (1976 est.) Ségou (36,400)
Economy: Based mainly upon farming and livestock raising
Chief Products: Agriculture—cotton, livestock, millet, peanuts, rice, shea nuts, sorghum; fishing—fresh and dried fish; mining—gold, iron, salt
Money: Basic unit—franc
Climates: Steppe, desert, tropical wet and dry

Mali is a large, thinly populated country with few natural resources or industries. Most of its people are farmers or herders. A large majority live in rural areas, chiefly in the south. Much of northern Mali is desert.

More than half the people of Mali are black Africans who belong to various ethnic groups. Most of them farm or raise livestock in southern Mali. They grow cotton and food crops such as rice, corn, yams, manioc, and peanuts. They live in villages of round, mud-brick huts. In the north, Moors, Arabs, and Tuareg people roam the desert to find grazing land for their livestock. They live in camel-hair tents and have a diet that consists largely of millet, dates, and camel's milk. Fulani people live in the Niger River valley. Some raise cattle and live in low huts made of straw mats or tree branches. Others grow crops and live in mud-brick houses.

France ruled Mali for more than 60 years before it became independent. French is still Mali's official language. Only about 10 per cent of Mali's adults can read and write.

Malta

Malta (95) is an island nation in the Mediterranean Sea. It is composed of two main islands and several smaller islands.

Capital: Valletta
Official Languages: Maltese and English
Government: Republic—president
serves as head of state and is appointed by
country's one-house parliament; prime minister
is country's most powerful official and is assisted by a Cabinet
National Anthem: "Innu Malti" ("Maltese Anthem")
Flag: A silver replica of the George Cross, a British medal awarded to Malta for bravery in World War II, appears on a red and white field
Area: 122 sq. mi. (316 km^2); greatest distances—east-west, 22 mi. (35.2 km); north-south, 19 mi. (30.4 km); coastline—58 mi. (93 km)
Population: 1983 estimate—377,000; distribution—83 per cent urban, 17 per cent rural; density—3,090 persons per sq. mi. (1,193 persons per km^2)
Largest Cities: (1979 est.) Sliema (20,095), Birkirkara (16,832), Qormi (15,784), and Valletta (14,042)
Economy: Based mainly on shipbuilding and ship repairing, with very limited agriculture and tourism
Chief Products: Agriculture—barley, grapes, onions, potatoes, wheat; manufacturing and processing—beverages, processed food, shipbuilding and repair
Money: Basic unit—Maltese pound
Climate: Subtropical dry summer

Malta is a tiny nation with fine natural harbors. These, along with the country's strategic location, have accounted for Malta's military importance in past years. The Mediterranean fleets of the British navy and the North Atlantic Treaty Organization (NATO) were once headquartered in Malta. The former British naval dockyards are now used for shipbuilding.

Tourists come to enjoy the mild climate and to see Malta's fine Baroque and Renaissance art and architecture. Malta also has ruins and relics of great interest. Remains of late Stone Age and Bronze Age people have been found in limestone caverns in Malta. The Phoenicians colonized Malta around 1000 B.C., and ruins of their civilization still stand.

The two official languages of Malta are English and Maltese, a West Arabic dialect with some Italian words. Roman Catholicism is the state religion. Malta has about 120 Roman Catholic schools and about 80 private schools. All children from 6 to 14 must attend school.

Fewer than 20 per cent of Malta's people live in rural areas. Farmers raise small crops of barley, wheat, grapes, onions, potatoes, and citrus fruits in the rocky soil. Most of the country's food is imported.

Mauritania

Mauritania (96) is located in western Africa. It is bordered by Western Sahara, Algeria, Mali, and Senegal.

Capital: Nouakchott
Official Name: République Islamique de Mauritanie (Islamic Republic of Mauritania)
Official Language: French; (national language—Arabic)
Government: Military rule—top military leader serves as president and is assisted by two vice-presidents, both military officers; one vice-president serves as premier who heads a Cabinet which carries out government operations
Flag: Green with a yellow star and crescent in the center; the green color and the star and crescent stand for Mauritania's ties to Islam and northern Africa; the yellow stands for the country's ties to nations south of the Sahara
Area: 397,956 sq. mi. (1,030,700 km^2); greatest distances—north-south, 800 mi. (1,287 km); east-west, 780 mi. (1,255 km); coastline—414 mi. (666 km)
Population: 1983 estimate—1,775,000; distribution—64 per cent rural, 36 per cent urban; density—5 persons per sq. mi. (2 per km^2)
Largest Cities: (1976 census) Nouakchott (134,986—metro area) and Nouadhibou (21,961)
Economy: Based mainly on agriculture, with some small industry and mining
Chief Products: Agriculture—dates, gum arabic, livestock (cattle, sheep, goats), millet; mining—iron ore; fishing—ocean and freshwater fish
Money: Basic unit—ouguiya
Climates: Steppe, desert

Mauritania is an Islamic nation populated mainly by Arabic-speaking people called Moors. Black Africans form a large minority group. Most of the people live in the southern part of the country, where there is enough rainfall to support agriculture. The Sahara covers northern Mauritania, and the desolate land is broken by only a few fertile oases. Exports of iron ore are an important source of income for Mauritania.

Poor communication and transportation systems have prevented Mauritania from developing its economy. About 90 per cent of the people of Mauritania are farmers or herders. The Moors move from place to place in search of water and grazing land for their livestock. They live in camel-hair tents. Most of the blacks live in villages of round, mud-brick houses that stand along narrow, winding pathways. They grow millet, rice, corn, and beans for food. Many blacks work as government employees or as teachers. Only about 10 per cent of the children in Mauritania attend primary school, and roughly 20 students complete high school every year.

Mauritius

Mauritius (97) is an island nation located in the Indian Ocean, about 500 miles (800 kilometers) east of Madagascar.

Capital: Port Louis
Official Language: English
Government: Constitutional monarchy—a governor general, appointed by Great Britain, represents the British Crown; a premier runs the government; 70-member Legislative Assembly passes the country's laws; the majority party in the assembly chooses the premier
National Anthem: "Motherland"
Flag: Four horizontal stripes of red, blue, yellow, and green (top to bottom); red stands for the struggle for freedom, blue for the Indian Ocean, yellow for the light of independence shining over the island, and green for agriculture
Area: 790 sq. mi. (2,045 km^2); greatest distances—north-south, 38 mi. (61 km); east-west, 29 mi. (47 km); coastline—100 mi. (161 km)
Population: 1983 estimate—1,003,000; distribution—52 per cent urban, 48 per cent rural; density—1,269 persons per sq. mi. (490 per km^2)
Largest Cities: (1979 est.) Port Louis (144,412); (1978 est.) Beau Bassin (83,714) and Curepipe (54,356)
Economy: Based mainly upon the raising and processing of sugar and sugar products
Chief Product: Agriculture—sugar
Money: Basic unit—rupee
Climate: Tropical wet

Mauritius is a small nation with an economy that depends largely on sugar cane. Nearly all its exports are sugar or sugar products. Fields of sugar cane cover about half the country's land area, and about two-thirds of all its workers grow, harvest, or process sugar cane. Mauritius also produces some tea and tobacco, and the people grow vegetables in small gardens or between the rows of sugar cane. Most of the country's food, however, must be imported.

The people of Mauritius are descendants of European settlers, African slaves, Chinese traders, and Indian laborers and traders. About two-thirds of the people are Indians. Most of the rest are of mixed ancestry. Chinese and Europeans form small minorities. The diversity of the population is reflected in religion and language, for example. Hindu temples, Christian churches, Buddhist pagodas, and Muslim mosques are scattered throughout the country. English is the official language, but most of the people speak a French dialect called Creole.

Mexico

Mexico (98) is located in North
America. It is bordered by the
United States, Belize, and Guatemala.

Capital: Mexico City
Official Name: Estados Unidos Mexi-
canos (United Mexican States)
Official Language: Spanish
Government: Republic—based on a constitution
that provides for a president, a national legislature
called the Congress, and a Supreme Court
National Anthem: "Himno Nacional de México" ("National Anthem of Mexico")
Flag: Three vertical stripes of green, white, and red (left to right); the country's
coat of arms appears on the white stripe
Area: 758,136 sq. mi. (1,963,564 km^2), including 2,071 sq. mi. (5,364 km^2) of
outlying islands; greatest distances—north-south, 1,250 mi. (2,012 km); east-
west, 1,900 mi. (3,060 km); coastline—6,320 mi. (10,170 km)
Population: 1983 estimate—74,507,000; distribution—67 per cent urban, 33
per cent rural; density—98 persons per sq. mi. (38 per km^2)
Largest Cities: (1980 census) Mexico City (9,373,353) and Guadalajara
(1,725,000)
Economy: Based mainly on agriculture, with increasing manufacturing, trade,
and petroleum
Chief Products: Agriculture—alfalfa, beans, coffee, corn, cotton, fruits, hene-
quen, livestock, rice, sugar cane, tobacco, vegetables, wheat; fishing—aba-
lones, oysters, sardines, shrimp, tuna; forestry—chicle, ebony, mahogany,
pine, rosewood; manufacturing—cement, chemicals, clothing, fertilizers, iron
and steel, handicraft articles, household appliances, processed foods, wood
pulp and paper; mining—coal, copper, fluorspar, gold, iron ore, lead, manganese, natural gas, petroleum, silver, sulfur, tin, zinc
Money: Basic unit—peso
Climates: Subtropical dry summer, steppe, desert, highlands, tropical wet and
dry, tropical wet

Mexico is a nation shaped by two cultures. Both its ancient Indian heritage and a long period of Spanish rule have influenced the country's way
of life. Most Mexicans are mestizos, or persons of mixed white and Indian
ancestry. Nearly all of them speak Spanish and are Roman Catholics.

Agriculture employs more Mexican workers than any other economic
activity, but manufacturing and the petroleum industry are growing rapidly. Rapid economic development has enabled many Mexicans to have
modern housing and to eat and dress well. But more than a third of the
people are poor. About two-thirds of all Mexicans live in urban areas.

Monaco

Monaco (99) is located on the Mediterranean Sea and is bordered on three sides by southeastern France.

Capital: Monaco
Official Language: French
Government: Principality—ruled by a prince who represents the country in international affairs; a minister of state, who is French and is nominated by the French government, heads Monaco's government; three councilors responsible for finance, police and internal affairs, and public works assist the minister of state; 18-member National Council is the country's legislative body; under terms of a treaty with France, if Monaco's royal family has no male heirs, Monaco will come under French rule
Flag: Two horizontal stripes of red and white (top to bottom)
Area: 0.58 sq. mi. (1.49 km^2); greatest distances—east-west, 1.5 mi. (2.4 km); north-south, 1.75 mi. (2.8 km); coastline—3 mi. (5 km)
Population: 1983 estimate—27,000; distribution—100 per cent urban; density—46,993 persons per sq. mi. (18,121 persons per km^2)
Largest Cities: (1968 census) La Condamine (11,438) and Monte Carlo (9,948)
Interesting Sights: Palace of the Prince in Monaco; the Marine Museum and the Museum of Prehistoric Anthropology, both in Monaco; the Museum of Fine Arts in Monte Carlo; the Casino of Monte Carlo
Economy: Based mainly on tourism, with some small, local industries
Chief Products: Beer, candy, chemicals, dairy products
Money: Basic unit—French franc
Climate: Subtropical dry summer

Monaco is one of the tiny countries of the world, but its picturesque location on the French Riviera and its many tourist attractions make it a popular resort area. Monaco has many fine beaches, hotels, and clubs, as well as museums, libraries, botanical gardens, a zoo, and a theater and orchestra. It is famous for its annual automobile races and for the Monte Carlo gambling casino.

French is the official language of Monaco, and its citizens are called Monégasques. Only about a seventh of the people are Monégasque, however. More than half are French, and most of the others are Americans, Belgians, British, or Italians. Many wealthy people from other countries live in Monaco because of the country's low tax rates. Many foreign corporations have headquarters there for the same reason. Monaco relies on tourism for most of its income. It also earns money selling colorful postage stamps.

Mongolia

Mongolia (100) is located in east-central Asia. It is bordered by Russia and China.

Capital: Ulan Bator
Official Name: Mongolian People's Republic
Official Language: Mongolian
Government: People's Democracy (Communist)—country's elected legislature, the Great National Khural, elects a seven-member Presidium, the nation's real governing body; Presidium's chairman is the head of state; Khural also elects a Council of Ministers which runs the government
Flag: Vertical stripes of red, blue, and red, with gold symbols on the left stripe; red stands for Communism and blue for the Mongols of the past
Area: 604,250 sq. mi. (1,565,000 km^2); greatest distances—east-west, 1,500 mi. (2,414 km); north-south, 790 mi. (1,271 km)
Population: 1983 estimate—1,795,000; distribution—50 per cent rural, 50 per cent urban; density—3 persons per sq. mi. (1 per km^2)
Largest City: (1980 census) Ulan Bator (418,700)
Economy: Based mainly on livestock raising and related industries, with some developing industry
Chief Products: Agriculture—camels, cattle, goats, grain, horses, meat, milk, potatoes, sheep, vegetables; manufacturing and processing—building materials, felt, processed foods, soap, textiles; mining—coal, petroleum
Money: Basic unit—tughrik
Climates: Subarctic, steppe, desert

Mongolia is a sparsely populated country of high plateaus, rugged mountains, and bleak deserts. Some Mongolians follow a way of life that has remained unchanged for hundreds of years. They live in collapsible felt tents and wander from place to place with herds of livestock. But more than half the people live and work on huge cooperative farms set up by the government. Sheep, camels, cattle, goats, and horses are raised on most of the farms. On others, the people grow grain and other crops. The farms are like huge ranches with small towns in the center. The central areas have houses, shops, offices, and medical facilities.

Almost all Mongolians are an Asian people called Mongols. The Mongolian language is written with a special form of the Cyrillic alphabet, the alphabet used by Russians.

Mongolia exports cattle, wool, dairy products, furs, hides, and meat. Russia is Mongolia's chief trade partner. A railroad links Mongolia with Russia in the north and China in the south.

Morocco

Morocco (101) is located in northwest Africa. It is bordered by Algeria and Western Sahara.

Capital: Rabat
Official Language: Arabic
Government: Constitutional monarchy—king is the head of state and appoints prime minister and other ministers to assist him; king presides over Council of Ministers, or Cabinet; 240-member Chamber of Representatives is Morocco's one-house legislature
National Anthem: "Al Nachid Al Watani" ("The National Anthem")
Flag: A green star centered on a red field
Area: 172,414 sq. mi. (446,550 km^2); greatest distances—east-west, 760 mi. (1,223 km); north-south, 437 mi. (703 km); coastline—Atlantic, 612 mi. (985 km); Mediterranean, 234 mi. (377 km)
Population: 1983 estimate—22,119,000; distribution—59 per cent rural, 41 per cent urban; density—129 persons per sq. mi. (50 persons per km^2)
Largest Cities: (1971 census) Casablanca (1,506,373), Rabat (367,620), Marrakech (332,741), and Fez (325,327)
Economy: Mainly agriculture and commerce
Chief Products: Agriculture—almonds, barley, beans, citrus fruits, corn, oats, olives, peas, wheat; manufacturing and processing—candles, cement and other building materials, foodstuffs, leather, soap, textiles; mining—clay, lead, limestone, marble, phosphates
Money: Basic unit—dirham
Climates: Subtropical dry summer, steppe, desert

Morocco is a small, mountainous country populated chiefly by Arabs and Berbers. The Arabs speak Arabic, Morocco's official language. The Berbers speak various Berber dialects.

Morocco has various minerals and ranks as the world's leading exporter of phosphates. But most Moroccans are farmers or herders. Many of the herders live in tents. A typical farmhouse is made of adobe or branches and has two rooms: one for the father and one for the mother and children. The people sleep on mats on the floor and eat their meals seated on the floor around a low table. *Pastilla,* a traditional dish, is a pie made with lamb, pigeon, chicken, eggs, vegetables, and spices.

Most Moroccan cities have a modern section with tall apartment buildings, lovely parks, and wide streets. In the older sections, old adobe buildings crowd narrow, winding streets, and merchants sell food, clothing, and other goods from open-air stalls. Roads and railroads link Morocco's large cities. Many people travel by mule or horse.

Mozambique

Mozambique (102) is located on the southeastern coast of Africa. It is bordered by Swaziland, South Africa, Zimbabwe, Zambia, Malawi, and Tanzania.

Capital: Maputo
Official Name: República Popular de Moçambique (People's Republic of Mozambique)
Official Language: Portuguese
Government: Communist dictatorship—controlled by the nation's only political party, the Front for the Liberation of Mozambique, known as Frelimo; president of Frelimo is also the nation's president; the party's 15-member Central Committee has highest governing power; the party appoints the 210-member legislature, the People's Assembly
Flag: Four wedge-shaped diagonal stripes of green, red, black, and yellow, which are separated by white bands; in the upper left corner, a white cogwheel encloses a book, which has a gun and a hoe crossed over it
Area: 309,496 sq. mi. (801,590 km^2); greatest distances—north-south, 1,100 mi. (1,770 km); east-west, 680 mi. (1,094 km); coastline—1,556 mi. (2,504 km)
Population: 1983 estimate—12,910,000; distribution—91 per cent rural, 9 per cent urban; density—41 persons per sq. mi. (16 per km^2)
Largest Cities: (1970 census) Maputo (341,922), Nampula (120,188), and Beira (110,752)
Economy: Based on agriculture and trade, with some limited industry
Chief Products: Cashews, coconuts, cotton, sugar
Money: Basic unit—Escudo
Climates: Subtropical moist, steppe, tropical wet and dry

Mozambique is a developing nation that relies mainly on agriculture to support its people. About 90 per cent of all Mozambicans live in rural areas. Almost all of them are black Africans who speak various Bantu languages. Portuguese is the country's official language, but most of the people cannot speak it.

The government owns all the farmland and the major industries in Mozambique. Most farmers use old-fashioned methods to raise crops. Cashew trees and coconut palms grow throughout Mozambique, producing the country's leading products. Coal mines and a few small industries employ some workers. Many others work in mines in neighboring South Africa.

Mozambique has fine natural harbors, excellent port facilities, and several railroads. It earns money from fees paid by neighboring countries for the use of its railroads and port facilities.

Nauru

Nauru (103) is an island nation located in the central Pacific Ocean, just south of the equator.

Capital: None
Official Languages: English and Nauruan
Government: Republic—18-member elected Parliament makes the country's laws; Parliament elects a president who selects a Cabinet; president and Cabinet carry out government operations
Flag: A horizontal gold stripe crosses a field of royal blue; below the stripe is a white 12-pointed star
Area: 8 sq. mi. (21 km²); greatest distances—east-west, 4 mi. (6.4 km); north-south, 3.5 mi. (5.6 km); coastline—12 mi. (19 km)
Population: 1983 estimate—8,000; distribution—100 per cent rural; density—987 persons per sq. mi. (381 persons per km²)
Largest City: None; population concentrated in coastal settlements and villages
Economy: Based almost entirely on the production of phosphates for export, with an attempt at developing a shipping industry
Chief Product: Phosphates
Money: Basic unit—Australian dollar
Climate: Tropical wet

Nauru is an oval-shaped coral island that is rich in phosphates—valuable chemicals used in making fertilizers. Phosphates are Nauru's only important resource and its only export. Income from phosphate exports has provided money to build homes, schools, and hospitals. The government is also saving some of the income to support the island's population after all the phosphates have been mined.

About half of Nauru's people are Nauruans—people of mixed Polynesian, Micronesian, and Melanesian ancestry. Most of the others are temporary residents who come to work in the phosphate mines. The Nauruans are Christians, and most speak English and the Nauruan language. Most of the men work in the phosphate industry.

The government provides the people with modern, low-rent houses and free medical care. Nauruan children between the ages of 6 and 17 must attend school. The government pays the expenses of students who attend universities in other countries.

Nepal

Nepal (104) is located in south Asia. It is bordered by China and India.

Capital: Kathmandu
Official Language: Nepali
Government: Constitutional monarchy—king serves as head of state and commander in chief of armed forces; the legislative council, the National Panchayat, makes the country's laws; the king can veto any law passed by the Panchayat or put into effect any law that it fails to pass; king selects all seven members of the Supreme Court and can reverse any of its decisions; many Nepalese believe the king is descended from the Hindu god Vishnu
National Anthem: "Rashtriya Dhun" ("National Anthem")
Flag: Two crimson triangles trimmed in blue, one above the other; the top triangle features the moon and the lower one the sun, symbols of the long life of Nepal; it is the only nonrectangular country flag
Area: 54,362 sq. mi. (140,797 km^2); greatest distances—east-west, 500 mi. (805 km); north-south, 150 mi. (241 km)
Population: 1983 estimate—14,955,000; distribution—95 per cent rural, 5 per cent urban; density—275 persons per sq. mi. (106 per km^2)
Largest City: (1971 census) Kathmandu (150,402)
Economy: Based mainly on agriculture, with some limited local industry
Chief Products: Cattle, corn, rice, oilseeds, wheat
Money: Basic unit—rupee
Climates: Highlands, tropical wet and dry

Nepal is covered largely by the highest mountains in the world—the Himalayas. Almost all its people are farmers or herders. In the mountains, where the climate is cold and harsh, the people herd sheep and yaks. Farmers in the foothills and valleys, where the climate is milder, raise crops and tend livestock. South of the foothills is a region of jungles and swamps, where animals such as crocodiles, elephants, leopards, rhinoceroses, and tigers live. A flat strip of fertile land lies along Nepal's southern border. Farmers there raise rice, sugar cane, tobacco, and other crops in the warm, rainy climate.

Nepal's people belong to various ethnic groups. Only about half speak Nepali, the official language. Over 30 other languages and dialects are spoken in Nepal. Hinduism is the state religion, but most Nepalese combine Hindu and Buddhist beliefs. Illiteracy is a major problem in Nepal. Only about 15 per cent of the people over the age of 6 can read and write. Nepalese soldiers, called Gurkhas, became known for their bravery while serving in the British and Indian armies.

112

The Netherlands

The Netherlands (105) is located in northwestern Europe. It is bordered by West Germany and Belgium.

Capital: Amsterdam
Also Called: Holland
Official Language: Dutch
Government: Constitutional monarchy—king or queen serves as head of state but with little real power; monarch names all appointed government officials; prime minister heads government assisted by a Cabinet; two-house parliament, called the States-General, passes laws
National Anthem: "Wilhelmus van Nassouwe" ("William of Nassau")
Flag: Three horizontal stripes of red, white, and blue (top to bottom)
Area: 15,892 sq. mi. (41,160 km^2), including 1,175 sq. mi. (3,043 km^2) of inland water; greatest distances—north-south, 196 mi. (315 km); east-west, 167 mi. (269 km); coastline—228 mi. (367 km)
Population: 1983 estimate—14,427,000; distribution—88 per cent urban, 12 per cent rural; density—909 persons per sq. mi. (351 persons per km^2)
Largest Cities: (1975 est.) Amsterdam (757,958), Rotterdam (620,867), and The Hague (482,879)
Economy: Based mainly on manufacturing and commerce
Chief Products: Agriculture—barley, dairy products, flower bulbs, oats, potatoes, sugar beets, wheat; manufacturing—clothing, electronic equipment, iron and steel, machinery, petroleum products, processed foods, textiles, transportation equipment; mining—natural gas, petroleum, salt
Money: Basic unit—guilder
Climate: Oceanic moist

The Netherlands is a small, densely populated country. Over two-fifths of its land was once under water. But the water has been pumped out, and the country's most fertile farmland and its largest cities lie on this reclaimed land. The people of The Netherlands, known as the Dutch, are protective of their land and take great care to keep their homes, towns, and countryside clean and neat.

The many rivers and canals in The Netherlands provide important transportation routes. When it is cold enough, the people also use the canals for one of their favorite sports: ice skating.

The Netherlands has a thriving economy that produces a variety of goods. But it is particularly well known for products such as cheese, chocolate, and flower bulbs. The Netherlands is also famous for its masters of painting, including Rembrandt and Vincent Van Gogh.

New Zealand

New Zealand (106) is an island nation located in the southwest Pacific Ocean. It is composed of two main islands and many smaller islands.

Capital: Wellington
Official Language: English
Government: Constitutional monarchy—governor general represents the British monarch in New Zealand, but has little real power; an elected, one-house Parliament, a prime minister, and an appointed Cabinet run the government
Anthems: "God Defend New Zealand" (national); "God Save the Queen" (royal) ·
Flag: Royal blue field with the British Union Flag in the upper left corner; to the right, four red stars outlined in white
Area: 103,883 sq. mi. (269,057 km^2); greatest distances—east-west, 600 mi. (960 km); north-south, 910 mi. (1,456 km)
Population: 1983 estimate—3,180,000; distribution—84 per cent urban, 16 per cent rural; density—31 persons per sq. mi. (12 per km^2)
Largest Cities: (1981 census) Auckland (766,183), Wellington (319,615), Christchurch (289,392), and Manukau (158,208)
Economy: Based mainly on manufacturing, farming, and trade
Chief Products: Agriculture—butter, cheese, meat, wool; manufacturing—chemicals, machinery, paper and wood pulp, petroleum products, plastics, processed foods, textiles, transportation equipment
Money: Basic unit—New Zealand dollar
Climates: Oceanic moist, subtropical moist

New Zealand is a beautiful country of snow-capped mountains, rolling green hills, and many lakes and waterfalls. Much of the land is used as pasture for millions of sheep and cattle. Livestock provides New Zealand with most of its income. The country exports large quantities of lamb, wool, butter, and cheese.

Most New Zealanders are descendants of British and other European settlers. Maoris, a Polynesian people whose ancestors were the first people to live in New Zealand, make up about 9 per cent of the population. New Zealand has a strong tradition of equal rights for all people. Its citizens also enjoy one of the world's high standards of living. They eat well, and most live in single-family homes with a small yard or garden. Almost every family has a car. The government provides excellent free health care to all New Zealanders and free education to students up to age 19.

Nicaragua

Nicaragua (107) is located in Central America. It is bordered by Honduras and Costa Rica.

Capital: Managua
Official Name: República de Nicaragua (Republic of Nicaragua)
Official Language: Spanish
Government: Republic (Junta)—Since May 1982, Nicaragua has been under a state of emergency. The country is ruled by a three-person Junta, which is assisted by a nine-person Directorate. Government policies are made by the Directorate and are subject to the approval of the Junta.
National Anthem: "Himno Nacional de Nicaragua" ("National Hymn of Nicaragua")
Flag: Two blue horizontal stripes separated by a white horizontal stripe; the nation's coat of arms lies in the center of the white stripe
Area: 50,193 sq. mi. (130,000 km²); greatest distances—north-south, 293 mi. (472 km); east-west, 297 mi. (478 km); coastlines—Pacific, 215 mi. (346 km); Caribbean, 297 mi. (478 km)
Population: 1983 estimate—2,980,000; distribution—53 per cent urban, 47 per cent rural; density—60 persons per sq. mi. (23 per km²)
Largest Cities: (1980 est.) Managua (677,680) and (1979 est.) León (83,693)
Economy: Based mainly on agriculture, with growing industry and commerce
Chief Products: Agriculture—bananas, beans, beef cattle, coffee, corn, cotton, rice, sesame, sugar cane; manufacturing—clothing and textiles, processed foods and beverages
Money: Basic unit—córdoba
Climates: Tropical wet and dry, tropical wet

Nicaragua is a developing nation with an economy that depends largely on agricultural exports. Cotton is the chief export, but Nicaragua also produces coffee, sugar cane, rice, bananas, beef, and sesame for export.

Most Nicaraguans have mixed Spanish and Indian ancestry. They speak Spanish and are Roman Catholics. Small minority groups of Indians live in thinly populated regions far from any cities or towns. They speak their own languages and follow traditional Indian ways of life.

Many Nicaraguan farmers work as laborers on large estates. Others own their own small farms. In the warm regions near the Pacific coast, farm families live in houses built of branches or poles, with palm-leaf roofs. Inland, where the climate is cooler, some people live in adobe houses with tile roofs. Only about half the children of Nicaragua attend school. Many rural areas have no schools. The country has about 9,000 miles (14,000 kilometers) of roads, but most are unpaved.

Niger

Niger (108) is located in west-central Africa. It is bordered by Algeria, Libya, Chad, Nigeria, Benin, Upper Volta, and Mali.

Capital: Niamey
Official Name: République du Niger (Republic of Niger)
Official Language: French
Government: Republic (military rule)—president serves as head of state and is also president of the Supreme Military Council, a Cabinet composed of both military and civilian members
Flag: Three horizontal stripes of orange, white, and green (top to bottom); an orange circle, representing the sun, lies in the center; orange represents the Sahara, white stands for purity, and green for agriculture
Area: 489,191 sq. mi. (1,267,000 km^2); greatest distances—east-west, 1,100 mi. (1,770 km); north-south, 825 mi. (1,328 km)
Population: 1983 estimate—5,780,000; distribution—87 per cent rural, 13 per cent urban; density—13 persons per sq. mi. (5 persons per km^2)
Largest Cities: (1977 est.) Niamey (225,300), Zinder (58,400), Maradi (45,900), and Tahoua (31,300)
Economy: Based almost entirely on agriculture
Chief Products: Agriculture—beans, chilies, cotton, henna, hides and skins, livestock (cattle, goats, sheep), manioc, millet, okra, onions, peanuts, peas, rice, sorghum; mining—natron, salt, tin, uranium
Money: Basic unit—franc
Climates: Steppe, desert

Niger is a thinly populated, landlocked nation. Much of the country is desert, and less than a fourth of the land can be used for farming. But most of the people are farmers or herders who produce just enough to feed themselves. Niger has few manufacturing industries, but it does have deposits of uranium and some other minerals.

Niger was once a French territory, and French is the official language. Few of the people speak French, however. The population consists of four ethnic groups, each with its own language. Fulani and Tuareg people live in northern and central Niger, where many of them roam from place to place with herds of livestock. They live on animal products including blood taken from living animals, meat, milk, and milk products. Djerma and Hausa people live in the south, where many are farmers who grow crops such as cotton, beans, millet, rice, and peanuts. About 80 per cent of the people of Niger are Muslims. Others practice fetish religions, in which they worship objects or images.

116

Nigeria

Nigeria (109) is located on the west coast of Africa. It is bordered by Niger, Chad, Cameroon, and Benin.

Capital: Lagos
Official Name: Federal Republic of Nigeria
Official Language: English
Government: Republic—elected president, assisted by an appointed Cabinet, carries out government operations; a two-house legislature, consisting of a Senate and House of Representatives, makes the country's laws
National Anthem: "Nigeria, We Hail Thee"
Flag: Two green vertical stripes, representing agriculture, separated by a white vertical stripe, symbolizing unity and peace
Area: 356,669 sq. mi. (923,768 km^2); greatest distances—east-west, 800 mi. (1,287 km); north-south, 650 mi. (1,046 km); coastline—478 mi. (769 km)
Population: 1983 estimate—84,721,000; distribution—80 per cent rural, 20 per cent urban; density—238 persons per sq. mi. (92 per km^2)
Largest Cities: (1977 est.) Lagos (1,149,200) and Ibadan (855,300)
Economy: Based mainly upon agriculture and the production of oil
Chief Products: Agriculture—beans, cacao beans, cassava, corn, cotton, livestock, millet, palm oil and palm kernels, peanuts, rice, rubber, yams; mining—columbite, limestone, petroleum, tin; manufacturing—cement, chemicals, clothing, food products, textiles
Money: Basic unit—Naira
Climates: Steppe, tropical wet and dry, tropical wet

Nigeria has more people than any other African nation. Its population consists of black Africans who belong to more than 250 different ethnic groups. Most Nigerians earn a living by farming, fishing, or herding. Nigeria is a leading producer of agricultural products such as cacao and peanuts. But rich petroleum deposits bring Nigeria most of its wealth and have made it one of the world's leading producers and exporters of oil.

Each of Nigeria's ethnic groups speaks its own language. English is the country's official language and is taught in most schools. About half the people are Muslims, and about a third are Christians.

Most Nigerians live in rural areas, where typical homes are made of grass, dried mud, or wood. Nigeria's large, crowded cities have modern houses and apartment buildings. Most rural people and many city dwellers wear traditional garments, including long, loose robes made of white or brightly colored fabrics. People throughout Nigeria enjoy the country's rich heritage of folk songs, dances, and stories.

Norway

Norway (110) is located in northern Europe. It is bordered by Sweden, Finland, and Russia.

Capital: Oslo
Official Name: Kongeriket Norge (Kingdom of Norway)
Official Language: Norwegian (Bokmål and Nynorsk)
Government: Constitutional monarchy—king is head of state, but with little real power; prime minister is head of government and selects a Cabinet to run government departments; elected Parliament, called the Storting, consists of one house, but its members form two sections to discuss and vote on laws
National Anthem: "Ja, vi elsker dette landet" ("Yes, We Love This Land")
Flag: A blue cross edged in white appears on a red field
Area: 125,182 sq. mi. (324,219 km²); greatest distances—northeast-southwest, 1,100 mi. (1,770 km); northwest-southeast, 280 mi. (451 km); coastline—1,650 mi. (2,655 km)
Population: 1983 estimate—4,128,000; distribution—56 per cent rural, 44 per cent urban; density—34 persons per sq. mi. (13 persons per km²)
Largest Cities: (1978 est.) Oslo (460,377) and Bergen (211,861)
Economy: Based largely upon manufacturing and trade
Chief Products: Agriculture—barley, dairy products, hay, livestock, oats, potatoes; fishing—capelin, cod, herring, mackerel; forestry—timber; manufacturing—aluminum, chemicals, processed foods, ships, wood pulp and paper; mining—ilmenite, iron ore, lead, molybdenite, petroleum, pyrites, zinc
Money: Basic unit—krone
Climates: Oceanic moist, highlands

Norway is a mountainous land, covered largely by bare rock. Only 3 per cent of its area is farmland, but its waters are a valuable resource. Norway's rugged coastline is marked by many long inlets of the sea called fiords, which provide excellent natural harbors. The coastal waters have abundant fish. Norway's shipping and fishing industries are among the world's largest. In addition, Norway gets petroleum from oil fields in the North Sea. Rivers that rush down Norway's mountains provide hydroelectric power to fuel its many manufacturing industries.

Norway's coastal regions, where most of the people live, have a somewhat mild climate. But the inland areas are often snow covered. For thousands of years, the people of Norway have used skis to travel about.

The government provides free health care and many welfare services. About 96 per cent of the people belong to the Evangelical Lutheran Church, Norway's official church.

Oman

Oman (111) is located on the southern tip of the Arabian Peninsula in Southwest Asia. It is bordered by the United Arab Emirates, Yemen (Aden), and Saudi Arabia.

Capital: Muscat
Official Language: Arabic
Government: Sultanate—a sultan acts as ruler and governs the country assisted by a five-member, appointed council; some people living in the mountains and other inland areas support the imam, an Islamic religious leader, instead
Flag: A red vertical stripe appears at the left, with white, red, and green horizontal stripes at the right; the national coat of arms appears in white silhouette in the upper left corner
Area: 82,030 sq. mi. (212,457 km²); greatest distances—north-south, 500 mi. (805 km); east-west, 400 mi. (644 km); coastline—about 1,200 mi. (1,930 km)
Population: 1983 estimate—976,000; distribution—93 per cent rural, 7 per cent urban; density—13 persons per sq. mi. (5 persons per km²)
Largest Cities: (1974 est.) Matrah (20,000), Salalah (10,000), and Nazwa (10,000)
Economy: Based mainly on agriculture and the oil industry
Chief Products: Agriculture—coconuts, dates, limes, pomegranates; mining—petroleum
Money: Basic unit—rial
Climate: Desert

Oman is a small, thinly populated country where the temperature sometimes reaches 130°F. (54°C). It has little manufacturing, few roads, and no railroads. But it is an important oil-producing nation.

Most of Oman's people are Arabs. The country has minorities of blacks, Indians, and Baluchis, or people whose ancestors came from the Baluchistan region of Pakistan. The Shuhuhs are a people who live in caves in the northern tip of Oman and exist mainly on fish from the Gulf of Oman.

Oman's inland areas consist mainly of barren plateaus and desert. Fertile areas along the coast produce coconuts, dates, limes, and pomegranates. Most Omanis farm or work in the petroleum industry. Some fish for a living or work for cattle and camel breeders. Most Omanis live in tents or in houses built of mud and stone. They dress in traditional long, loose garments and headdresses that protect them from the sun and sand.

119

Pakistan

Pakistan (112) is located in southern Asia. It is bordered by Iran, Afghanistan, China, and India.

Capital: Islamabad
Official Name: The Islamic Republic of Pakistan
Official Language: Urdu
Government: Military rule—military leaders took control of the government in 1977; Army Chief of Staff declared himself president, dissolved the Parliament, and suspended the Constitution; appointed Cabinet assists the president; an appointed Federal Advisory Council replaced the Parliament but without legislative powers
National Anthem: "Qaumi Tarana" ("National Anthem")
Flag: One white vertical stripe at the left; the rest of the flag is green with a white star and crescent, the traditional symbols of Islam
Area: 310,404 sq. mi. (803,943 km^2); greatest distances—north-south, 935 mi. (1,505 km); east-west, 800 mi. (1,287 km); coastline—506 mi. (814 km)
Population: 1983 estimate—89,230,000; distribution—72 per cent rural, 28 per cent urban; density—287 persons per sq. mi. (111 persons per km^2)
Largest Cities: (1972 census) Karachi (3,515,402), Lahore (2,169,742), Lyallpur (823,343), Hyderabad (628,631), and Rawalpindi (614,809)
Economy: Based mainly on agriculture, with developing manufacturing
Chief Products: Agriculture—barley, cotton, fruits, oilseeds, rice, sugar cane, tobacco, wheat; manufacturing—cement, cotton textiles, fertilizer; mining—coal, limestone, natural gas, petroleum
Money: Basic unit—Pakistani rupee; one hundred paisas equal one rupee
Climates: Steppe, desert, highlands

Pakistan is a Muslim nation populated by various ethnic groups, each with its own language. Other differences divide the people, but Islam is an important link. Daily prayers and other Muslim rituals are an important part of life in Pakistan.

About three-fourths of all Pakistanis live in rural areas, and two-thirds of all the country's workers are farmers or herders. Many of them use simple tools to grow wheat and other crops on small plots of land. The people also eat rice, fruits, vegetables, and some meat. Many of the people live in simply furnished two- or three-room houses made of clay or sun-dried mud. Most rural houses have no plumbing or electricity.

Pakistan has a shortage of schools and teachers, and only about a fifth of the people can read and write. Paved roads and railroads link many cities and towns in Pakistan. But in rural areas, the villagers use camels, cattle, donkeys, or horses for transportation.

Panama

Panama (113) is located in Central America. It is bordered by Colombia and Costa Rica.

Capital: Panama City
Official Name: República de Panamá (Republic of Panama)
Also Called: Crossroads of the World
Official Language: Spanish
Government: Republic—president heads the government assisted by a Cabinet; the National Assembly of Community Representatives is elected by the people and serves as the nation's lawmaking body
National Anthem: "Himno Nacional de la República de Panamá" ("National Hymn of the Republic of Panama")
Flag: Divided into quarters which are (clockwise from top left) white, red, white, blue; a blue star, for honesty and purity, appears in the upper white segment; a red star, for authority and law, appears in the lower white segment
Area: 29,856 sq. mi. (77,326 km^2); greatest distances—east-west, 410 mi. (660 km); north-south, 130 mi. (209 km); coastline—Atlantic Ocean, 397 mi. (639 km); Pacific Ocean, 746 mi. (1,201 km)
Population: 1983 estimate—2,013,000; distribution—51 per cent urban, 49 per cent rural; density—67 persons per sq. mi. (26 per km^2)
Largest Cities: (1980 est.) Panama City (465,160) and San Miguelito (169,870)
Economy: Based mainly upon agriculture and transportation
Chief Products: Agriculture—bananas, beans, cocoa, coffee, corn, rice, sugar cane; manufacturing—cement, clothing, petroleum products, processed foods; fishing—shrimp; forestry—mahogany; mining—copper, gold
Money: Basic unit—balboa
Climates: Tropical wet and dry, tropical wet

Panama is a small nation with worldwide importance as a transportation center. The Panama Canal, which links the Atlantic and Pacific oceans, cuts through Panama. Thousands of ships pass through the canal each year, giving Panama its nickname, *Crossroads of the World.*

Most Panamanians are of mixed white and American Indian ancestry or mixed white and black ancestry. About 95 per cent of the people are Roman Catholics. Nearly all speak Spanish.

Almost all Panamanians live near the canal or west of it. Eastern Panama consists largely of swamps and jungles. The canal area is a busy urban center, where many people work in commerce, trade, manufacturing, banking, and businesses related to the canal. Western Panama is an agricultural region of tiny farming villages and towns.

Papua New Guinea

Papua New Guinea (114) occupies the eastern half of the island of New Guinea and a chain of smaller islands. It is bordered by Indonesia.

Capital: Port Moresby
Official Language: English
Government: Constitutional monarchy—British monarch serves as head of state and is represented on the islands by a governor general; prime minister heads the government and is selected by an elected national legislature
Flag: Divided diagonally from upper left to lower right; a golden bird of paradise is in the upper section, which is red; five stars representing the Southern Cross appear in the lower section, which is black
Total Land Area: 178,260 sq. mi. (461,691 km^2); greatest distances between islands—north-south, 730 mi. (1,174 km); east-west, 1,040 mi. (1,674 km)
Population: 1983 estimate—3,348,000; distribution—87 per cent rural, 13 per cent urban; density—18 persons per sq. mi. (7 per km^2)
Largest Cities: (1977 est.) Port Moresby (106,600) and Lae (45,100)
Economy: Based almost entirely upon agriculture, with some limited manufacturing and mining
Chief Products: Agriculture—cocoa, coconuts, coffee, rubber, tea, timber; mining—copper, gold, silver
Money: Basic unit—kina
Climates: Tropical wet and dry, tropical wet

Papua New Guinea is a nation populated almost entirely by dark-skinned people called Melanesians. More than 700 languages are spoken in the country. The people use several widely understood languages to communicate with one another. Fewer than 30 per cent of the people have received any elementary education.

Tropical forests cover most of Papua New Guinea, and many of the coastal areas are swamps. Houses in the swamplands are built on stilts to keep them cool and protect them from moisture. Farming is the chief occupation in Papua New Guinea. The people grow most of their own food, including sweet potatoes, yams, cassava, and taro plants. They also grow cocoa, coconuts, coffee, tea, and other crops for export. The country's most important export is copper.

Papua New Guinea has one daily newspaper and about 15 radio stations. There are about 10,000 miles (16,000 kilometers) of roads in the country, but most are unpaved. A national airline provides service between the islands of Papua New Guinea.

Paraguay

Paraguay (115) is located in South America. It is bordered by Brazil, Argentina, and Bolivia.

Capital: Asunción
Official Name: República del Paraguay (Republic of Paraguay)
Official Language: Spanish
Government: Republic—elected president is the nation's strongest political leader; president selects a Cabinet of five or more ministers to help carry out government operations; president also receives advice from the Council of State; country has an elected, two-house legislature
National Anthem: "Himno Nacional del Paraguay" ("National Anthem of Paraguay")
Flag: Three horizontal stripes of red, white, and blue (top to bottom), believed to honor French ideals; national coat of arms is centered on the front of the flag; the treasury seal with a lion and liberty cap appears on the back
Area: 157,048 sq. mi. (406,752 km^2); greatest distances—north-south, 575 mi. (925 km); east-west, 410 mi. (660 km)
Population: 1983 estimate—3,361,000; distribution—60 per cent rural, 40 per cent urban; density—21 persons per sq. mi. (8 persons per km^2)
Largest City: (1972 census) Asunción (392,753)
Economy: Based mainly on agriculture, with some manufacturing
Chief Products: Agriculture—cattle, citrus fruits, corn, cotton, rice, sugar cane, tobacco; forest products—lumber, petitgrain oil, quebracho extract, yerba maté; manufacturing and processing—canned meats, leather goods, processed fruits, vegetable oils
Money: Basic unit—guaraní; one hundred céntimos are equal to one guaraní
Climates: Subtropical moist, steppe, tropical wet and dry

Paraguay is a small nation with an economy that depends mainly on agriculture and forest products. Paraguay has fertile soil, but only about 3 per cent of the land is cultivated. Most farmers grow just enough to feed themselves. Some also raise cattle. Cotton and tobacco are the chief export crops. Two others come from trees. Quebracho trees provide tannin, which is used to tan leather. The bitter orange tree produces an oil used in marmalade and perfume.

Most Paraguayans are of mixed white and Indian ancestry, and nearly all are Roman Catholics. Guaraní Indians form a tiny minority group. Spanish is Paraguay's official language, but Guaraní is widely spoken.

Most Paraguayans earn little money, and they live in small homes with few modern conveniences. Religious holidays are important social events, often celebrated with festivals. Soccer is a popular sport.

123

Peru

Peru (116) is located on the west coast of South America. It is bordered by Ecuador, Colombia, Brazil, Bolivia, and Chile.

Capital: Lima
Official Name: República del Perú (Republic of Peru)
Official Languages: Spanish and Quechua
Government: Republic—elected president heads the government aided by two elected vice-presidents; the two-house legislature, consisting of a Senate and a Chamber of Deputies, is also elected and makes the country's laws
National Anthem: "Himno Nacional del Perú" ("National Hymn of Peru")
Flag: Two vertical red stripes separated by a vertical white stripe; a shield appears in the center of the white stripe
Area: 496,225 sq. mi. (1,285,216 km^2); greatest distances—north-south, 1,225 mi. (1,971 km); east-west, 875 mi. (1,408 km); coastline—1,448 mi. (2,330 km)
Population: 1983 estimate—19,317,000; distribution—63 per cent urban, 37 per cent rural; density—39 persons per sq. mi. (15 per km^2)
Largest Cities: (1972 census) Lima (2,941,473), Callao (313,316), and Arequipa (302,316)
Economy: Based largely on manufacturing, mining, and agriculture
Chief Products: Agriculture—bananas, coffee, cotton, potatoes, sugar cane; fishing—anchovettas; manufacturing—fish meal, metals, sugar, textiles; mining—copper, iron ore, lead, petroleum, silver, zinc
Money: Basic unit—sol
Climates: Desert, highlands, tropical wet

Peru has the largest Indian population of any country in the Western Hemisphere. Nearly half its people are Indians. Most of the rest are of mixed white and Indian ancestry. Peru's Indians are descendants of the Inca and other tribes. Spain conquered the great Inca empire in the 1500's and made Peru a colony. Indian and Spanish influences are both evident in present-day Peru.

Peru ranks among the world's leaders in fishing and in the production of copper, lead, silver, and zinc. It has a small minority of wealthy people and a growing middle class, but most Peruvians are poor. Farming is the chief occupation.

About 95 per cent of the people are Roman Catholics, though some Indian Catholics still worship Inca gods. Religious holidays are often celebrated with parades, feasts, games, and dancing.

124

Philippines

The Philippines (117) is an island nation located off the coast of Southeast Asia.

Capital: Manila
Official Name: Republic of the Philippines
Official Languages: Pilipino and English
Government: Parliamentary republic—elected president heads the government and appoints a prime minister who directs the government's day-to-day operations; the prime minister appoints a 20-member Cabinet; the country's legislature is called the National Assembly
National Anthem: "Lupang Hinirang" ("Land That I Love")
Flag: Two horizontal stripes of blue and red meet a white triangle at the mast; three yellow stars and a sun appear in the white triangle
Area: 115,831 sq. mi. (300,000 km²); greatest distances—north-south, 1,152 mi. (1,854 km); east-west, 688 mi. (1,107 km); coastline—8,052 mi. (12,958 km)
Population: 1983 estimate—51,598,000; distribution—64 per cent rural, 36 per cent urban; density—445 persons per sq. mi. (172 per km²)
Largest Cities: (1975 census) Manila (1,479,116) and Quezon City (956,864)
Economy: Mainly agricultural, with some mining and growing manufacturing
Chief Products: Agriculture—abacá, bananas, coconuts, corn, pineapples, rice, sugar cane, tobacco; forestry—ebony, kapok, Philippine mahogany; fishing industry—fish, shellfish, mother-of-pearl, sponges; mining—chromite, cinnabar, copper, gold, iron ore, limestone, manganese, nickel, silver; manufacturing—cement, chemicals, clothing, foods, petroleum products, textiles
Money: Basic unit—peso
Climates: Tropical wet and dry, tropical wet

The Philippines has a variety of natural resources, including fertile soil. Agriculture is the leading occupation. The country is an important producer of coconuts and sugar. Rice is the chief food crop. Forests in the Philippines have over 3,000 kinds of trees that provide lumber and other products. Philippine mines produce gold and other minerals. The waters surrounding the islands provide abundant fish.

People live on about 900 of the more than 7,000 islands that make up the Philippines. Ships are an important means of transportation between the islands. The country also has good roads and railroads.

The people of the Philippines speak many languages and dialects. The schools teach English as well as Pilipino, the national language. The Philippines has more Christians than any other country in Asia.

125

Poland

Poland (118) is located in central Europe. It is bordered by Russia, Czechoslovakia, and East Germany.

Capital: Warsaw
Official Name: Polska Rzeczypospol-
ita Ludowa (Polish People's Republic)
Official Language: Polish
Government: People's republic (Communist dictatorship)—government is centered
around a one-house, 460-member legislature
called the Sejm; Sejm passes laws and supervises other government
branches; 17 members of the Sejm serve on the Council of State; the Sejm
also appoints a Council of Ministers, including the prime minister; Communists hold the majority of seats in these bodies
National Anthem: "Jeszcze Polska nie Zginęła" ("Poland Has Not Yet Per-
ished")
Flag: Two horizontal stripes of white and red (top to bottom); nation's coat of
arms appears on the white stripe
Area: 120,725 sq. mi. (312,677 km²); greatest distances—east-west, 430 mi.
(692 km); north-south, 395 mi. (636 km); coastline—277 mi. (446 km)
Population: 1983 estimate—36,463,000; distribution—58 per cent urban, 42
per cent rural; density—303 persons per sq. mi. (117 per km²)
Largest Cities: (1978 est.) Warsaw (1,552,400), Łódź (825,200), Kraków
(693,200), and Wrocław (597,700)
Economy: Based mainly upon manufacturing and heavy industries
Chief Products: Agriculture—barley, hogs, potatoes, rye, sugar beets, wheat;
manufacturing—chemicals, food products, iron and steel, machinery, ships;
mining—coal, zinc
Money: Basic unit—zloty
Climate: Continental moist

Poland is a leading industrial nation of Eastern Europe. Polish factories produce iron and steel, machinery, ships, chemicals, and other products. About 85 per cent of the country's farmland is privately owned. Collective farms and government-owned state farms occupy the rest of the land.

Nearly 60 per cent of all Poles live in urban areas. Most city dwellers live in small two- or three-room apartments. In rural areas, many people live in brick or wooden cottages. About half of all Polish families own a radio and television set. Only about 2 per cent own a car.

Most Poles are Roman Catholics. For many people, social life centers around family gatherings and church events. Poland has a rich cultural heritage of art, literature, and music.

126

Portugal

Portugal (119) is located in western Europe. It is bordered by Spain.

Capital: Lisbon
Official Name: República Portuguesa
(Portuguese Republic)
Official Language: Portuguese
Government: Republic—elected president appoints a prime minister;
Cabinet assists prime minister in carrying out government operations; 250-member Parliament makes country's laws
National Anthem: "A Portuguesa" ("The Portuguese")
Flag: A green vertical stripe at the mast; the rest of the flag is red; the country's coat of arms lies in the center
Area: 35,553 sq. mi. (92,082 km²), including the Azores and Madeira island groups; mainland, excluding the islands, 34,340 sq. mi. (88,941 km²); greatest distances; mainland—north-south, 350 mi. (563 km); east-west, 125 mi. (201 km); coastline, mainland—458 mi. (737 km)
Population: 1983 estimate—10,264,000; distribution—69 per cent rural, 31 per cent urban; density—287 persons per sq. mi. (111 per km²)
Largest Cities: (1973 est.) Lisbon (757,700) and Porto (304,000)
Economy: Mainly mining and manufacturing, with agriculture and fishing
Chief Products: Agriculture—almonds, corn, figs, grapes, lemons, limes, olives, oranges, rice, wheat; fishing—cod, sardines, tuna; manufacturing—clothing, cork products, food products, leather goods, metals and machinery, petroleum products, ships, textiles
Money: Basic unit—escudo
Climate: Subtropical dry summer

Portugal is famous for its historic role as the center of the vast colonial empire it established in the 1400's and 1500's. Portugal no longer rules a great empire. Today, its people generally have a lower standard of living than most other Europeans. The majority of Portuguese live in small farming and fishing villages. Portuguese fishermen use rowboats to fish in rough Atlantic waters. Women and children clean the fish and mend the nets. Farmers grow crops including citrus fruits, olives, and grapes, which are made into fine wines.

In Portugal's fast-growing cities, many people work in manufacturing industries. The cities have modern apartment and office buildings, as well as buildings that are hundreds of years old. There are magnificent churches from Portugal's golden age of art in the 1400's and 1500's. Most Portuguese are Roman Catholics, and their religion is an important influence on everyday life.

Qatar

Qatar (120) is located in south-
western Asia, on a peninsula that
reaches into the Persian Gulf. It
is bordered by Saudi Arabia and the
United Arab Emirates.

Capital: Doha
Official Name: The State of Qatar
Official Language: Arabic
Government: Emirate—ruled by an emir who appoints a 14-member Council of Ministers to as-
  sist him; an advisory council of 20 elected deputies and 3 deputies appointed
  by the Council of Ministers also aids the emir; the government does not
  allow political parties; the emir is a member of the ruling al-Thani family,
  which took control of the country in the early 1800's
Flag: The left half is white and adjoins the jagged edge of the right half which is
  maroon
Area: 4,247 sq. mi. (11,000 km²); greatest distances—north-south, 115 mi. (185
  km); east-west, 55 mi. (89 km); coastline—235 mi. (378 km)
Population: 1983 estimate—242,000; distribution—79 per cent urban, 21 per
  cent rural; density—57 persons per sq. mi. (22 persons per km²)
Largest City: (1975 est.) Doha (140,000)
Economy: Based mainly upon the oil industry, with very limited agriculture
Chief Products: Petroleum and petroleum products, fertilizer, cement
Money: Basic unit—Qatar Dubai riyal; 100 dirhams equal one riyal
Climate: Desert

Qatar is a small Arab nation with an economy that depends almost entirely on the oil industry. Exports of petroleum and petroleum products
account for about 95 per cent of the country's income.

Most of the people of Qatar are Arabs. Many came from other Arab
countries and settled in Qatar after the development of the oil industry in
the 1950's. Islam is the state religion of Qatar. Arabic is the official language, and many of the people also speak English. Most Qataris live in
or near the capital city of Doha. They have modern housing, and the
government provides free education and health care. Children from the
ages of 6 to 16 must attend school. Many older Qataris cannot read or
write. Special schools have helped raise the adult literacy rate.

Most of Qatar is covered by stony desert and barren salt flats. Irrigation
has helped develop some agriculture. Qatar produces enough vegetables
for its people, but it must import meat and other foods. Qatar has a few
manufacturing industries and a shrimp-fishing fleet.

Romania

Romania (121) is located in eastern Europe. It is bordered by Russia, Bulgaria, Yugoslavia, and Hungary.

Capital: Bucharest
Official Name: Republica Socialistă România (Socialist Republic of Romania)
Official Language: Romanian
Government: Socialist republic (Communist dictatorship)—Communist Party leaders hold all political power; government exists to carry out the party's policies; the party's Standing Presidium is the real policymaking body of both the party and the country; the party's head, called the general secretary, is chairman of the Standing Presidium and has the authority of a dictator; people elect a one-house legislature, called the Grand National Assembly, which meets to approve laws made by the Standing Presidium
National Anthem: "Imnul de Stat al Republicii Socialiste România" ("State Anthem of the Socialist Republic of Romania")
Flag: Three vertical stripes of blue, yellow, and red (left to right), which are the national colors; the coat of arms appears in the center of the flag
Area: 91,699 sq. mi. (237,500 km^2); greatest distances—east-west, about 450 mi. (724 km); north-south, about 320 mi. (515 km)
Population: 1983 estimate—22,874,000; distribution—51 per cent rural, 49 per cent urban; density—249 persons per sq. mi. (96 persons per km^2)
Largest Cities: (1977 census) Bucharest (1,807,044) and Cluj (262,421)
Economy: Based mainly upon manufacturing, mining, and agriculture
Chief Products: Agriculture—corn, fruits, potatoes, sheep, wheat; industry—clothing, food, iron ore, machinery, natural gas, petroleum
Money: Basic unit—leu
Climate: Continental moist

Romania is a nation with rich farmland, vast forests, and large mineral deposits. But Romanians generally have a lower standard of living than most other Europeans. They have adequate food, clothing, and shelter but few can afford luxury items. Less than 1 per cent of the people own a car. Buses provide the chief means of transportation in the cities.

About half of all Romanians live in rural areas, where small wooden cottages provide housing. On special occasions, the rural people wear colorful costumes and enjoy folk music and dancing. In the cities, most people live in crowded apartments. The cities offer various forms of entertainment, including concerts, movies, and plays.

The government owns or controls all Romanian farms and factories. Heavy machinery is Romania's chief manufactured product.

Russia

Russia (122) is located in northern Eurasia. It is bordered by Norway, Finland, Poland, Czechoslovakia, Hungary, Romania, Turkey, Iran, Afghanistan, China, Mongolia, and North Korea.

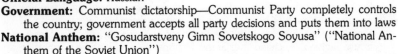

Capital: Moscow
Official Name: Union of Soviet Socialist Republics (U.S.S.R.)
Also Called: The Soviet Union
Official Language: Russian
Government: Communist dictatorship—Communist Party completely controls the country; government accepts all party decisions and puts them into laws
National Anthem: "Gosudarstveny Gimn Sovetskogo Soyusa" ("National Anthem of the Soviet Union")
Flag: A red field with a gold hammer and sickle in upper left corner; a gold star for the Communist Party appears above the hammer and sickle
Area: 8,649,500 sq. mi. (22,402,000 km²); greatest distances—east-west, 6,000 mi. (9,656 km); north-south, 3,200 mi. (5,150 km)
Population: 1983 estimate—272,775,000; distribution—62 per cent urban, 38 per cent rural; density—31 persons per sq. mi. (12 persons per km²)
Largest Cities: (1979 census) Moscow (7,831,000) and Leningrad (4,073,000)
Economy: Mainly agriculture and industry
Chief Products: Agriculture—barley, cattle, corn, milk, oats, potatoes, rye, sheep, sugar beets, wheat; fishing—cod, herring, salmon, sturgeon; manufacturing—chemicals, electronic equipment, iron and steel, lumber, machinery, paper, petroleum products, processed foods, processed metals, textiles; mining—bauxite, coal, copper, gold, iron ore, lead, natural gas, nickel, petroleum, platinum, salt, tungsten, zinc
Money: Basic unit—ruble
Climates: Extremely varied—polar, subarctic, continental moist, subtropical moist, steppe, desert, highlands

In area, Russia is the largest country in the world. It has more farmland, forests, and mineral deposits than any other nation and ranks second only to the United States in the total value of its production. Russians are known for their outstanding contributions to literature, music, and the performing arts. Russia's people belong to more than 100 nationality groups, each with its own language. Russian is the official language.

The government controls the economy, the education system, and many other areas of Russian life. Nearly two-thirds of all Russians live in urban areas, where typical housing consists of small apartments. In rural villages, many Russians live in log houses or community barracks.

Rwanda

Rwanda (123) is located in central Africa. It is bordered by Uganda, Tanzania, Burundi, and Zaire.

Capital: Kigali
Official Name: Republic of Rwanda
Also Called: The African Switzerland
Official Languages: French and Kinyarwanda
Government: Republic—headed by an elected president who appoints a 16-member Cabinet to assist him; an elected one-house legislature, called the National Development Council, approves policies and laws proposed by the president
Flag: Three vertical stripes of red, yellow, and green (left to right); a large black *R* appears in the center of the yellow stripe
Area: 10,169 sq. mi. (26,338 km^2); greatest distances—east-west, 145 mi. (233 km); north-south, 110 mi. (177 km)
Population: 1983 estimate—5,450,000; distribution—96 per cent rural, 4 per cent urban; density—536 persons per sq. mi. (207 persons per km^2)
Largest Cities: (1978 census) Kigali (117,700), Butare (21,700), and Ruhengeri (16,000)
Economy: Based mainly upon agriculture, with some limited mining
Chief Products: Agriculture—cattle, coffee, pyrethrum, tea; mining—tin, wolfram
Money: Basic unit—franc
Climate: Tropical wet and dry

Rwanda is a small, densely populated country with beautiful lakes and snow-capped mountains. But it is one of the poor nations of Africa. It has no railroads, little industry, and poor soil. Most of its people are farmers who grow barely enough to feed themselves. Some also grow coffee, which is Rwanda's chief export. European companies mine tin and wolfram, which are also exported.

More than 95 per cent of Rwanda's people live in rural areas. Most of the people belong to the Bahutu ethnic group. The Bahutu are farmers. Watusi people make up the chief minority group. They raise cattle and live on beef, milk, milk products, and blood drawn from living animals. Some make their clothing from animal hides and hair.

Rwanda's official languages are French and Kinyarwanda, a Bantu language spoken by most of the people. About half the people practice traditional African religions. Most of the others are Roman Catholics. Churches operate most of Rwanda's elementary schools, but there is a shortage of schools. About 75 per cent of the people cannot read or write.

131

Saint Lucia

St. Lucia (124) is a small island country in the West Indies. It lies in the Caribbean Sea about 240 miles (386 kilometers) north of Venezuela.

Capital: Castries
Official Language: English
Government: Constitutional monarchy—governor general is appointed by and represents the British monarch; prime minister, however, heads the government aided by a 10-member Cabinet; a two-house legislature passes the nation's laws
National Anthem: "Sons and Daughters of Saint Lucia"
Flag: Three triangles rising from a common base—first one is yellow; behind that, black; and finally, white, all on a blue field; flag represents the nation's volcanic peaks rising from the golden sands in the middle of the blue sea
Area: 238 sq. mi. (616 km^2); greatest distances—east-west, 14 mi. (22.4 km); north-south, 28 mi. (44.8 km)
Population: 1983 estimate—121,000; distribution—52 per cent rural, 48 per cent urban; density—508 persons per sq. mi. (196 per km^2)
Largest City: (1979 est.) Castries (45,000)
Economy: Mainly agricultural
Chief Products: Agriculture—bananas, coconuts; manufacturing—clothing, electrical parts, paper products, textiles
Money: Basic unit—East Caribbean dollar
Climate: Tropical wet

St. Lucia has an economy based on agriculture. The people grow crops primarily for their own use as food. Bananas and coconuts are exported. The country's few manufacturing industries make clothing, electrical parts, paper products, and textiles.

About 90 per cent of the people of St. Lucia are descendants of black African slaves brought to the island by early British and French settlers. Whites make up most of the rest of the population. Many have British or French ancestry. English is the country's official language and is widely spoken, but many St. Lucians speak a French dialect. More than 90 per cent of the people are Roman Catholics.

A large majority of the people of St. Lucia live in rural areas. Palm trees and other tropical vegetation cover much of the land. A paved road encircles the island and links the main towns with the capital city of Castries. Wooden houses painted in pastel colors are characteristic dwellings of St. Lucia.

Saint Vincent and the Grenadines

St. Vincent and the Grenadines (125) is a small island country in the West Indies. It lies about 200 miles north of Venezuela.

Capital: Kingstown
Official Language: English
Government: Constitutional monarchy—governor general is a symbolic official appointed by the British monarch; prime minister, the head of the political party with the most seats in Parliament, runs the government assisted by a Cabinet; a one-house Parliament, consisting of 13 elected representatives and 6 appointed senators, makes the country's laws
Flag: Three horizontal stripes of blue, yellow, and green (left to right) each separated by a narrow band of white; nation's coat of arms lies in the center of the yellow stripe
Area: 150 sq. mi. (388 km^2); greatest distances—north-south, 55 mi. (88 km); east-west, 25 mi. (40 km); coastline—32 mi. (84 km)
Population: 1983 estimate—126,000; distribution—52 per cent rural, 48 per cent urban; density—842 persons per sq. mi. (325 per km^2)
Largest City: (1977 est.) Kingstown (29,831)
Economy: Mainly agricultural, with some fishing, manufacturing, and tourism
Chief Products: Agriculture—arrowroot, bananas, coconuts, spices, plantains, peanuts, cotton, sugar, copra, root crops; manufacturing and industry—tourism, food products
Money: Basic unit—East Caribbean dollar
Climate: Tropical wet

Volcanic eruptions formed the 100 mountainous islands that make up St. Vincent and the Grenadines. Mount Soufrière, an active volcano on the northern end of St. Vincent, is the country's highest point. It rises 4,048 feet (1,234 meters) above the Caribbean Sea.

Most of the people of St. Vincent and the Grenadines are descendants of black African slaves brought to the islands by British and French settlers. English is the official national language, but many of the people also speak French. They live in wood or concrete houses with tile roofs. Their diet includes bananas, fish, rice, and a special dish of baked breadfruit and fried fish.

Though quite small, this country is a major world supplier of one agricultural product: arrowroot. The plant's roots are made into starch. The word *arrowroot* goes back to the Arawak Indians, the islands' first inhabitants. Their word, *araruta,* meant "mealy root."

San Marino

San Marino (126) is the smallest republic in Europe. It is located in the Apennine Mountains, and it is entirely surrounded by Italy.

Capital: San Marino

Official Name: La Serenissima Repubblica di San Marino (The Most Serene Republic of San Marino)

Official Language: Italian

Government: Republic—elected legislature, called the Grand and General Council, makes the country's laws; council selects two members to head the government; these two captains-regent appoint 10 government department heads; the heads of all San Marino families meet twice a year to discuss public matters

Flag: A blue and a white horizontal stripe; the San Marinese coat of arms is in the center and features the three towers on Mt. Titano

Area: 24 sq. mi. (61 km^2); greatest distances—north-south, 7 mi. (11.2 km); east-west, 6 mi. (9.6 km)

Population: 1983 estimate—22,000; distribution—74 per cent urban, 26 per cent rural; density—935 persons per sq. mi. (360 per km^2)

Largest City: (1977 est.) San Marino (4,628)

Interesting Sights: Spectacular mountain views; fortress wall and its three towers surrounding the capital city of San Marino; old churches, houses, and winding cobblestone streets in San Marino

Economy: Based upon agriculture, tourism, and limited local manufacturing

Chief Products: Animal hides, barley, chestnuts, fruit, wheat, building stone, ceramics, leather goods, textiles, tiles, varnish

Money: Basic unit—Italian lira

Climate: Subtropical dry summer

San Marino is the smallest republic in Europe and one of the smallest countries in the world. It was founded in the A.D. 300's. Its historic architecture and scenic location on the slopes of the Apennine Mountains attract more than 2½ million tourists a year. The country's historic buildings include a church built in the 1300's, which stands on a hill overlooking the walled city of San Marino, the nation's capital.

Tourism is an important source of income for San Marino. The country also has some industry and earns money from the sale of colorful postage stamps, which collectors in many parts of the world buy.

San Marino is completely surrounded by Italy, and its people speak Italian. They have much in common with the northern Italians, but the people of San Marino are proud of their independence and traditions.

São Tomé and Príncipe

São Tomé and Príncipe (127) is composed of two large islands and several smaller ones that lie about 180 miles (290 kilometers) off the west coast of Africa.

Capital: São Tomé
Official Language: Portuguese
Government: Republic—people elect a national assembly; assembly chooses a president who appoints a prime minister and a Cabinet to assist him in running the government
Flag: Green horizontal stripes (for forests and the sea) at the top and bottom; a yellow horizontal stripe (for soil) in the center; and a red triangle (for the struggle for freedom) near the staff; two black stars symbolizing the country's two main islands appear on the yellow stripe
Total Land Area: 372 sq. mi. (964 km^2); greatest distances—north-south, 120 mi. (192 km); east-west, 70 mi. (112 km); coastline—98 mi. (158 km)
Population: 1983 estimate—88,000; distribution—83 per cent rural, 17 per cent urban; density—236 persons per sq. mi. (91 per km^2)
Largest City: (1970 census) São Tomé (17,380)
Economy: Based almost entirely on agriculture and fishing
Chief Products: Cocoa, coconuts, coffee, copra, livestock
Money: Basic unit—escudo
Climate: Tropical wet and dry

São Tomé and Príncipe is an African island republic in the Gulf of Guinea. Most of the country's people live in rural areas and work on farms. By law, children are required to complete elementary school, but many do not.

Formerly a Portuguese colony, São Tomé and Príncipe has three population groups. About 70 per cent of the people have a mixed black African and European ancestry. They are called Creoles. Africans from mainland countries form the second large population group. Europeans account for a small percentage of the population. Roman Catholicism is the main religion among the Creoles and Europeans. The mainland Africans practice the religion of the country of their origin.

About 90 per cent of the nation's cultivated land belongs to large agricultural companies. The other 10 per cent is divided among about 11,000 small farm owners. The country's new leaders have announced plans to distribute farmland among more people.

Saudi Arabia

Saudi Arabia (128) is located in southwestern Asia. It is bordered by Jordan, Iraq, Kuwait, Qatar, United Arab Emirates, Oman, Yemen (Sana), and Yemen (Aden).

Capital: Riyadh

Official Name: Al-Mamlaka Al-Arab-iyya Al-Saudiyya (Kingdom of Saudi Arabia)

Official Language: Arabic

Government: Monarchy—based on the laws of Islam; king is both the chief political leader and the supreme religious leader; leading members of the royal family, which consists of several thousand persons, select a king from among themselves; king appoints a Council of Ministers to assist him in managing the government; king himself serves as prime minister

National Anthem: "Al-Salaam Al-Malaka As-Saudi" ("Royal Anthem of Saudi Arabia")

Flag: A white sword beneath a Muslim religious inscription written in Arabic, both centered on a field of green

Area: 830,000 sq. mi. (2,149,690 km^2); greatest distances—north-south, 1,145 mi. (1,843 km); east-west, 1,290 mi. (2,076 km); coastline—1,174 mi. (1,889 km) on the Red Sea; 341 mi. (549 km) on the Persian Gulf

Population: 1983 estimate—9,169,000; distribution—67 per cent urban, 33 per cent rural; density—10 persons per sq. mi. (4 per km^2)

Largest Cities: (1974 census) Riyadh (666,840), Juddah (561,104), Mecca (366,801), and At-Tā'if (204,857)

Economy: Based mainly upon the oil industry and agriculture

Chief Products: Agriculture—camels, citrus fruits, dates, goats, rice, vegetables, wheat; manufacturing—cement, fertilizer, food products; mining—petroleum

Money: Basic unit—riyal

Climates: Steppe, desert

Saudi Arabia is a large Middle Eastern monarchy. Beneath its vast deserts lie some of the world's large oil deposits. The income from oil sales is bringing modernization to this once-poor country. A rapid increase in the urban population has come about since the mid-1900's. Many Saudis, however, still live in rural areas. Most are farmers or nomads.

Nearly all Saudi Arabians are Arab Muslims. The country is specially honored in the Muslim world. Mecca and Medina, the two holiest cities of Islam, are located in Saudi Arabia. Islam influences family relationships, education, and many other aspects of everyday life. Most Saudis recite prayers five times daily.

Senegal

Senegal (129) is in west Africa. It is bordered by Mauritania, Mali, Gambia, Guinea, and Guinea-Bissau.

Capital: Dakar
Official Name: Republic of Senegal
Official Language: French
Government: Republic—elected president serves as head of state and chief executive; president appoints a prime minister; 100-member, elected legislature, called the National Assembly, makes the country's laws
Flag: Three vertical stripes of green, gold, and red (left to right); a green star lies in the center of the gold stripe
Area: 75,750 sq. mi. (196,192 km²); greatest distances—north-south, 300 mi. (480 km); east-west, 400 mi. (640 km); coastline—310 mi. (499 km)
Population: 1983 estimate—6,114,000; distribution—68 per cent rural, 32 per cent urban; density—80 persons per sq. mi. (31 per km²)
Largest Cities: (1976 census) Dakar (798,792), Thiès (117,333), and Kaolack (106,899)
Economy: Based mainly upon agriculture, with commerce and mining also important
Chief Products: Agriculture—beans, cassava, corn, livestock (cattle, goats, sheep), millet, peanuts, potatoes, rice, sorghum, sweet potatoes; mining—phosphates, zirconium; fishing—tuna; manufacturing and processing—beer, cement, cotton goods, peanut products
Money: Basic unit—franc
Climates: Steppe, tropical wet and dry

Senegal is an African republic that was once a French colony. Most of the people are black Africans who belong to various ethnic groups. About 8 out of 10 are Muslims, and many of the rest are Christians. Some practice traditional African religions.

Senegal has a rigid caste system that plays an important part in the lives of the people. The castes are (1) nobles; (2) freeborn; (3) artisans, or skilled workers; (4) griots, or musicians and praise singers; and (5) former slaves and their descendants. Ethnic groups often intermarry, but marriage between members of different castes is extremely rare.

Dakar, the capital of Senegal, is one of the important industrial centers in western Africa. The city is also a major seaport and an important stop on air routes between the United States and Africa and between Europe and South America.

Seychelles

Seychelles (130) is an island nation in the Indian Ocean about 1,000 miles (1,600 kilometers) east of the African mainland.

Capital: Victoria
Official Languages: English and French
Government: Republic—elected president heads the government and appoints a Cabinet to help carry out government operations; 23-member, elected legislature, called the Legislative Assembly, makes the country's laws
Flag: Top half is red; bottom half consists of a white horizontal stripe and a green horizontal stripe
Area: 171 sq. mi. (443 km^2); greatest distances—north-south, 260 mi. (416 km); east-west, 320 mi. (512 km); coastline—190 mi. (491 km)
Population: 1983 estimate—69,000; distribution—74 per cent rural, 26 per cent urban; density—404 persons per sq. mi. (156 per km^2)
Largest City: (1971 census) Victoria (13,736)
Economy: Based mainly on tourism, with a growing fishing industry
Chief Products: Agriculture—cinnamon, coconuts, copra breadfruit, bananas, cloves, tea, cassava, sweet potatoes; manufacturing and industry—processed copra, coconut oil, vanilla, tea; construction; tourism
Money: Basic unit—rupee
Climate: Tropical wet

Seychelles is a republic consisting of about 90 islands east of the African mainland. The country was ruled by Great Britain from 1814 until 1976. About 90 per cent of the people of Seychelles have mixed African and European ancestry. The others are Chinese, Indians, and Europeans of British or French origin.

About 35 per cent of the nation's workers are employed by the government. Another 25 per cent work in the construction industry, and about 15 per cent are farmers. A growing fishing industry contributes to the economy, but the main businesses are tourism and construction industry for tourism. The remote location and beautiful beaches attract many vacationers. Cinnamon grows wild on the island of Mahé, and coconut palms flourish on Mahé and many other islands. The country has numerous unusual species of plants and birds, and giant tortoises also live there.

Approximately 60 per cent of Seychelles' people can read and write. Most are Roman Catholics. A major problem in Seychelles is overpopulation.

Sierra Leone

Sierra Leone (131) is located in western Africa. It is bordered by Guinea and Liberia.

Capital: Freetown
Official Language: English
Government: Republic—president serves as head of government and is elected by the House of Representatives, the country's lawmaking body; 84 members of the House of Representatives are elected by the people and 12 members are chiefs of local ethnic groups
Flag: Three horizontal stripes of green, white, and blue (top to bottom)
Area: 27,699 sq. mi. (71,740 km^2); greatest distances—north-south, 220 mi. (354 km); east-west, 190 mi. (306 km); coastline—210 mi. (338 km)
Population: 1983 estimate—4,071,000; distribution—75 per cent rural, 25 per cent urban; density—148 persons per sq. mi. (57 per km^2)
Largest Cities: (1974 census) Freetown (274,000) and Bo (30,000)
Economy: Mainly agriculture and mining
Chief Products: Agriculture—bananas, cassava, cattle, cocoa, coffee, ginger, millet, palm products, peanuts, piassava, rice, sorghum; fishing—tuna; mining—chrome ore, diamonds, iron ore, rutile
Money: Basic unit—leone
Climates: Tropical wet and dry, tropical wet

Sierra Leone is a small country on Africa's western "bulge," north of the equator. A former British colony, the country became independent in 1961. A large portion of one of the world's valuable treasures—diamonds—comes from Sierra Leone. Diamonds make up about two-thirds of the total value of the country's exports.

Most of Sierra Leone's people are Africans of various ethnic groups. Education is developing rapidly there. Though English is the official language, most of the people speak African languages and practice local African religions. The Mende people, for example, believe that Ngewo, "God," created the world and everything in it. They worship small human images called *nomoli*, which they find in the soil. No one knows who made the figures. Other religions of Sierra Leone are Christianity and Islam.

Farmers of Sierra Leone produce a wide variety of crops including rice. But because of poor soil and the lack of modern farming methods, crop yields are low. Palm kernels are the chief export crop.

Singapore

Singapore (132) is an island nation that lies south of the Malay Peninsula in Southeast Asia.

Capital: Singapore
Official Name: Republic of Singapore
Official Languages: Chinese, English, Malay, and Tamil
Government: Republic—president serves as head of state, but with little real power; prime minister, head of the political party with the most seats in Parliament, runs the government assisted by an appointed Cabinet; one-house, elected Parliament makes the country's laws
National Anthem: "Majullah Singapura" ("Forward Singapore")
Flag: Two horizontal stripes, red on top (for equality and brotherhood) and white below (for purity and virtue); a white crescent and five white stars (for democracy, peace, progress, justice, and equality) lie in the upper left corner
Total Land Area: 238 sq. mi. (616 km^2); greatest distances (on Singapore island)—east-west, 26 mi. (42 km); north-south, 14 mi. (23 km); coastline—32 mi. (51 km)
Population: 1983 estimate—2,577,000; distribution—100 per cent urban; density—10,834 persons per sq. mi. (4,183 per km^2)
Largest City: (1980 est.) Singapore (2,390,800)
Economy: Mainly trade and commerce with manufacturing also important
Chief Products: Manufacturing and processing—chemicals, electronic equipment, machinery, metals, paper, petroleum products, processed food, rubber, ships, textiles, transportation equipment; agriculture—eggs, pork, poultry, fruits, vegetables
Money: Basic unit—Singapore dollar
Climate: Tropical wet

Singapore is an independent island country. Most of its people live in the city of Singapore, and the country's thriving economy is based on the city's businesses and banks. The 36-square-mile (93-square-kilometer) harbor boasts the best facilities in Southeast Asia. Singapore is the gateway to many places but especially to the Far East from India. Ships sailing from India to China, Japan, and Australia stop at Singapore to load, unload, and store cargo.

About 75 per cent of Singaporeans are Chinese. They run most of the import trade and much of the country's other businesses. About 15 out of 100 persons are Malays. Many of the Malays work in the fishing industry or as police officers, plantation laborers, or taxicab drivers. Most of the other people in the country are Indians and Europeans. The Europeans work mainly in trade and commerce.

Solomon Islands

Solomon Islands (133) is an island country in the South Pacific Ocean. It lies about 1,000 miles (1,610 kilometers) northeast of Australia.

Capital: Honiara
Official Language: English
Government: Constitutional monarchy—governor general represents the British monarch in the islands; prime minister, leader of the political party with the most seats in Parliament, runs the government aided by an eight-member, appointed Cabinet; 38-member, elected Parliament makes the country's laws
National Anthem: "God Save Our Solomon Islands"
Flag: Divided diagonally from lower left to upper right; upper left is blue with five white stars; lower right is green; a narrow band of yellow separates the two colors
Total Land Area: 11,500 sq. mi. (29,785 km²); greatest distances—north-south, 320 mi. (512 km); east-west, 500 mi. (800 km); coastline—2,051 mi. (5,313 km)
Population: 1983 estimate—241,000; distribution—91 per cent rural, 9 per cent urban
Largest City: (1976 census) Honiara (14,942)
Economy: Based upon agriculture, fishing, and forestry
Chief Products: Agriculture—coconuts, cocoa, copra, chilies, spices, taro; manufacturing and industry—fishing, forestry, food products, jute, rattan, asbestos fabrics, leather goods; gold panning
Money: Basic unit—Solomon Islands dollar
Climate: Tropical wet

Formerly a protectorate of Great Britain, the Solomon Islands gained independence in 1978. Its largest islands are Choiseul, Guadalcanal (the scene of fierce fighting between Allied and Japanese forces during World War II), Malaita, New Georgia, San Cristobal, and Santa Isabel.

Most Solomon Islanders are dark-skinned people called Melanesians. More than 90 per cent live in rural villages. Many people build houses on stilts to keep the dwellings cool. Their main foods include chicken, fish, pork, coconuts, and taro, a tropical plant with one or more edible, rootlike stems. English is the official language, but about 90 other languages are spoken among the Melanesians. The islanders also speak Pidgin English, which helps them cross language barriers. About 80 per cent of the people are Protestants.

141

Somalia

Somalia (134) is a nation located
in eastern Africa. It is bordered
by Djibouti, Kenya, and Ethiopia.

Capital: Mogadiscio
Formerly Called: British Somaliland
and Italian Somaliland
Official Language: Somali
Government: Military rule—military
officers hold top posts in both the government
and the country's only political party, the Somali Socialist
Revolutionary Party; head of the party serves as the country's president; a
Council of Ministers, appointed by party leaders, helps carry out government
operations; a People's Assembly approves government policies made by
party leaders; party leaders choose candidates for the assembly
Flag: Light blue with a large white star in the center
Area: 246,201 sq. mi. (637,657 km^2); greatest distances—north-south, 950 mi.
(1,529 km); east-west, 730 mi. (1,175 km); coastline—1,837 mi. (2,956 km)
Population: 1983 estimate—3,914,000; distribution—70 per cent rural, 30 per
cent urban; density—16 persons per sq. mi. (6 per km^2)
Largest City: (1977 est.) Mogadiscio (444,882)
Economy: Mainly agriculture, including livestock raising
Chief Products: Agriculture—bananas, grains, hides and skins, livestock, milk,
sugar; manufacturing—processed foods
Money: Basic unit—shilling
Climates: Steppe, desert, tropical wet and dry

Somalia is the easternmost country in Africa. It consists of two regions
that were once called British Somaliland and Italian Somaliland. These
colonies became independent in 1960 and combined to form the nation
of Somalia. Military leaders have controlled the country's government
since 1969 when Somalia's president was assassinated.

More than 70 per cent of the Somalis are nomads who raise herds of
camels, cattle, goats, and sheep. The others live on farms or in Somalia's
few towns. Almost all the people are Muslims. The nomads roam over
vast stretches of their hot, dry land in search of water and pasture for
their herds. They live in small, collapsible huts, and when time permits,
they enjoy telling stories and holding poetry contests in the evening
around a campfire. They recite long poems about major events such as
battles and victories, or in praise of prized possessions such as camels
and horses.

Somalia's chief exports are agricultural products. There is little manufacturing. Somalia has no railroads and only about 450 miles (724 kilometers) of roads are surfaced.

South Africa

South Africa (135) lies at the southern tip of Africa. It is bordered by Namibia, Botswana, Zimbabwe, Swaziland, Mozambique, and Lesotho.

Capitals: Cape Town (legislative); Pretoria (administrative); Bloemfontein (judicial)
Official Name: Republiek van Suid-Afrika (Republic of South Africa)
Official Languages: Afrikaans and English
Government: Republic—one-house Parliament, the House of Assembly, is elected by qualified voters; president serves as head of state and is elected by Parliament; prime minister, usually the leader of the majority party in Parliament, runs the government aided by a Cabinet
National Anthem: "Die Stem van Suid-Afrika" ("The Call of South Africa")
Flag: Three horizontal stripes of orange, white, and blue (top to bottom); small reproductions of the British Union Flag and the flags of the two former Boer republics appear centered on the white stripe
Area: 471,445 sq. mi. (1,221,037 km^2); greatest distances—east-west, 1,010 mi. (1,625 km); north-south, 875 mi. (1,408 km); coastline—about 1,650 mi. (2,655 km)
Population: 1983 estimate—31,444,000; distribution—50 per cent urban, 50 per cent rural; density—67 persons per sq. mi. (26 per km^2)
Largest Cities: (1970 census) Cape Town (691,296), Johannesburg (654,682), and Pretoria (543,950)
Economy: Mainly mining and manufacturing, with some agriculture
Chief Products: Agriculture—cattle, corn (maize), dairy products, fruits, sheep, sugar, wheat, wine, wool; manufacturing—chemicals, clothing, metals, metal products, processed foods; mining—asbestos, coal, copper, diamonds, gold, platinum, uranium
Money: Basic unit—rand
Climates: Subtropical moist, subtropical dry summer, steppe, desert

The people of South Africa form one of the world's complicated racial patterns: (1) The blacks, also called Africans, make up about 68 per cent of the population. (2) About 18 per cent are white. (3) About 11 per cent are called Colored people and are mixed black, white, and Asian. (4) Asians make up about 3 per cent of the population.

A government goal is the "separate development" of each racial group. This policy is called apartheid. Only whites may vote in parliamentary elections, serve in Parliament, and administer the laws.

South Africa has one of the world's strong economies. It has long been famous for its fabulous deposits of gold and diamonds.

Spain

Spain (136) is located in southwest Europe. It occupies most of the Iberian Peninsula. It is bordered by France, Andorra, and Portugal.

Capital: Madrid
Official Language: Castilian Spanish
Government: Parliamentary monarchy—king serves as head of state and acts as an advisor in matters of government policy; prime minister, head of the majority party in Parliament, serves as head of government and officially heads a Cabinet which carries out day-to-day government operations; two-house, elected Parliament, called the Cortes, makes the country's laws
National Anthem: "Himno Nacional" ("National Anthem")
Flag: A red horizontal stripe at the top and bottom; a wider yellow stripe in the center; the country's coat of arms appears on the yellow stripe
Area: 194,885 sq. mi. (504,750 km^2), including Balearic and Canary islands; greatest distances—east-west, 646 mi. (1,040 km); north-south, 547 mi. (880 km); coastline—2,345 mi. (3,774 km)
Population: 1983 estimate—38,679,000; distribution—74 per cent urban, 26 per cent rural; density—199 persons per sq. mi. (77 per km^2)
Largest Cities: (1975 est.) Madrid (3,201,234), Barcelona (1,754,714), Valencia (714,086), and Seville (590,235)
Economy: Mainly agriculture, services, and manufacturing
Chief Products: Agriculture—olives, oranges, wheat, wine grapes; manufacturing—automobiles, cement, chemical products, clothing, ships, steel
Money: Basic unit—peseta
Climates: Tropical dry summer, steppe, oceanic moist

During the 1950's and 1960's, Spain changed from a poor agricultural nation into a modern industrial one under the dictatorship of Francisco Franco, who had ruled Spain since 1939. Upon Franco's death in 1975, Spaniards began setting up a democratic government.

Almost all Spaniards are Roman Catholics. Celebrations to honor patron saints last several days. People decorate the streets, build bonfires, dance and sing, set off fireworks, and hold parades and bullfights.

Spain is noted for its great painters—Picasso among them. Spain's architecture—the Roman bridges, the mosques and palaces—shows the influence of various peoples who once controlled the country. As for other arts, folk singing and dancing have long been popular. Spanish dances such as the bolero, fandango, and flamenco have become world famous. Perhaps Spain is best known for its bullfighting. Most cities have at least one bull ring, and leading matadors are national heroes.

Sri Lanka

Sri Lanka (137) is an island nation located in the Indian Ocean off the southeastern tip of India.

Capital: Colombo
Official Name: Democratic Socialist Republic of Sri Lanka
Formerly Called: Ceylon
Official Languages: Sinhala and Tamil
Government: Republic—president heads the government assisted by a prime minister and Cabinet; 168-member legislature is called the National State Assembly
Flag: A gold lion on a crimson field to the right is a symbol of a precolonial Sri Lankan state; ornaments in the corners of the crimson field are gold bo leaves, or Buddhist symbols; two vertical stripes of green for Muslims and orange for Tamils appear at the left and represent the minorities; the entire flag is bordered in gold with a gold band separating the two parts
Area: 25,332 sq. mi. (65,610 km^2); greatest distances—north-south, 274 mi. (441 km); east-west, 142 mi. (229 km); coastline—725 mi. (1,167 km)
Population: 1983 estimate—15,640,000; distribution—73 per cent rural, 27 per cent urban; density—616 persons per sq. mi. (238 per km^2)
Largest Cities: (1977 est.) Colombo (616,000) and Dehiwala-Mount Lavinia (169,000)
Economy: Mainly agriculture and mining
Chief Products: Agriculture—cacao, cinnamon, coconuts, cotton, pepper, rice, rubber, tea, tobacco; mining—aquamarine, graphite, limestone, moonstone, ruby, sapphire, topaz, tourmaline, zircon
Money: Basic unit—rupee
Climates: Tropical wet and dry, tropical wet

For more than 2,000 years, Sri Lanka has been famous for its spices, tea and rubber plantations, and precious stones like sapphires and rubies. This beautiful island country was long a British colony. It became independent in 1948 and is now a republic with president, prime minister, cabinet, and legislature.

The country has six distinct groups of people. The Sinhalese, the largest group, are Buddhists whose predecessors came from northern India. The Tamils, descendants of people from southern Italy, are Hindus. The Moors, descendants of Arabs, are Muslims. The Malays are also Muslims but descend from people who originally came from Malaysia. The Burghers are descendants of European settlers who intermarried with Sri Lankans. Most of the Burghers are Christians and live in cities. The Veddahs, descendants of the country's first-known residents, live in remote forest regions and practice traditional local religions.

145

Sudan

Sudan (138) is located in northeastern Africa and is bordered by Egypt, Ethiopia, Kenya, Uganda, Zaire, Central African Republic, Chad, and Libya.

Capital: Khartoum
Official Name: Jumhuriyat as Sudan ad Dimuqratiyah (Democratic Republic of The Sudan)
Formerly Called: Anglo-Egyptian Sudan
Official Language: Arabic
Government: Republic—elected president heads the government and appoints a Cabinet to carry out government operations; elected legislature is called the People's Assembly
Flag: Three horizontal stripes of red, white, and black (top to bottom), with a green triangle near the staff symbolizing Islam
Area: 967,500 sq. mi. (2,505,813 km^2); greatest distances—north-south, 1,400 mi. (2,253 km); east-west, 1,075 mi. (1,730 km); coastline—400 mi. (644 km)
Population: 1983 estimate—20,246,000; distribution—75 per cent rural, 25 per cent urban; density—21 persons per sq. mi. (8 per km^2)
Largest Cities: (1973 census) Khartoum (333,921) and Omdurman (299,401)
Economy: Based mainly upon agriculture
Chief Products: Agriculture—cassava, corn, cotton, dates, hides and skins, melons, millet, peanuts, sesame, wheat; forest industry—gum arabic, hardwood; manufacturing and processing—beer, cement, salt, shoes, soap, textiles
Money: Basic unit—Sudanese pound
Climates: Steppe, desert, tropical wet and dry, tropical wet

The largest country in Africa, the Sudan sprawls from steaming tropical rain forest to grassy plains to bleak desert. It is in the rain forest where big game live, including lions, leopards, elephants, buffalo, giraffes, and the rare rhinoceros.

Most of the people of Sudan are farmers or herders who raise camels or cattle. About a third are African Negroes, who speak their own tribal languages and practice tribal religions. Arabic-speaking Muslims make up about two-thirds of the population. They are the descendants of African blacks and Nubians, who were brown-skinned people related to the early Egyptians and Libyans. These Muslims intermarried with Arab peoples and adopted their language and religion.

Sudan became independent in 1956. Since then, political unrest and civil war have plagued the government. It has changed hands between civilian and military leaders many times.

Suriname

Suriname (139) is located on the northeastern coast of South America. It is bordered by French Guiana, Brazil, and Guyana.

Capital: Paramaribo
Formerly Called: Dutch Guiana
Official Language: Dutch
Government: Republic—executive body, called the Policy Center, composed of military and civilian leaders heads the government; commander of the army ranks as most powerful member of the Policy Center; prime minister, who is also a member of the Policy Center, heads the Council of Ministers which carries out government operations
National Anthem: "Opo Kondre Man Oen Opo" ("Get Up People, Get Up")
Flag: Five horizontal stripes of green, white, red, white, and green (top to bottom); a yellow star lies in the center
Area: 63,037 sq. mi. (163,265 km^2); coastline—226 mi. (364 km); greatest distances—north-south, 285 mi. (459 km); east-west, 280 mi. (451 km)
Population: 1983 estimate—361,000; distribution—55 per cent rural, 45 per cent urban; density—5 persons per sq. mi. (2 per km^2)
Largest City: (1971 census) Paramaribo (102,300)
Economy: Based mainly upon mining and metal processing, with some agriculture
Chief Products: Aluminum, bananas, bauxite, rice
Money: Basic unit—guilder
Climate: Tropical wet

On the northeast coast of South America, mountainous rain forests cover most of Suriname. A majority of the people live in the flat coastal area. The Hindustanis own and operate small farms, and others are skilled industrial workers. Another group, the Creoles, work in government or for businesses. Many who are Indonesians are tenant farmers who rent their land. The black Africans, called Bush Negroes, descend from slaves who escaped in the 1600's. The Bush Negroes now live in the rain forests and follow African tribal customs. Each ethnic group has preserved its own culture, religion, and language.

The Netherlands ruled Suriname during most of the period from 1667 to 1975, when the country gained independence. About 70 per cent of Suriname's people can read and write. The country has one university which is located in Paramaribo.

The economy of Suriname is based primarily on mining and metal processing. Agriculture is important. Major crops include rice, bananas, coconuts, sugar, and lumber.

Swaziland

Swaziland (140) is in southern Africa. It is almost entirely surrounded by South Africa, except for a small area that borders Mozambique.

Capital: Mbabane (administrative); Lobamba (traditional)
Official Name: Kingdom of Swaziland
Official Languages: English and siSwati
Government: Monarchy—the Ngwenyama, or hereditary leader, rules the country as king assisted by a council of ministers and a legislature; the Ndlovukazi, or queen mother, is in charge of national rituals
Flag: Five horizontal stripes; the top and bottom stripes are blue (for peace); the wide center stripe is red (for past battles) with a black and white shield, spears, and staff; between the blue and red stripes are narrow yellow stripes (for natural resources)
Area: 6,704 sq. mi. (17,363 km^2); greatest distances—north-south, 120 mi. (193 km); east-west, 90 mi. (140 km)
Population: 1983 estimate—603,000; distribution—91 per cent rural, 9 per cent urban; density—91 persons per sq. mi. (35 per km^2)
Largest Cities: (1976 census) Mbabane (23,109) and Manzini (10,019)
Economy: Mainly agriculture and mining
Chief Products: Agriculture—corn, sugar cane, cotton, rice, tobacco, citrus fruits, hides and skins; manufacturing—cement, fertilizer, food products, wood products; mining—asbestos, iron ore
Money: Basic unit—lilangeni (plural spelled emalageni)
Climates: Subtropical moist, steppe

Beautiful little Swaziland has rich mineral deposits, large forests, and good farm and ranch land. White Europeans own almost all the mines, processing plants, and profitable farms. Ninety per cent of the population are Swazi, or black Africans. Most of them are peasant farmers. They are a proud, handsome, courteous people.

Formerly a British protectorate, Swaziland gained its independence in 1968. The Ngwenyama, or hereditary leader, rules the country as king.

Some of the Swazi live in towns and work in factories, offices, and shops; but most farm and raise livestock in extended family groups. Their homesteads once consisted of circular huts built around a cattle pen. Today, many Swazi live in Western-style houses. More than half the Swazi are Christians. Most of the rest practice a traditional African religion.

Sweden

Sweden (141) is a nation located in northern Europe on the Scandinavian Peninsula. It is bordered by Finland and Norway.

Capital: Stockholm
Official Name: Konungariket Sverige (Kingdom of Sweden)
Official Language: Swedish
Government: Constitutional monarchy—king serves as head of state, but his duties are largely ceremonial; prime minister and cabinet hold executive power; one-house, elected parliament is called the Riksdag
National Anthem: "Du gamla, du fria" ("Thou Ancient, Thou Free-Born")
Flag: One horizontal stripe and one vertical stripe, both of yellow, forming a cross on a blue field
Area: 173,732 sq. mi. (449,964 km²); greatest distances—north-south, 977 mi. (1,572 km); east-west, 310 mi. (499 km); coastline—4,700 mi. (7,564 km)
Population: 1983 estimate—8,370,000; distribution—87 per cent urban, 13 per cent rural; density—49 persons per sq. mi. (19 per km²)
Largest Cities: (1977 est.) Stockholm (658,435), Göteborg (440,082), and Malmö (238,454)
Economy: Based mainly upon manufacturing and various service industries
Chief Products: Agriculture—barley, livestock (cattle, hogs), milk and other dairy products, oats, potatoes, rye, sugar beets, wheat; fishing—cod, herring, mackerel, salmon; forestry—fir, pine, spruce; manufacturing—agricultural machinery, aircraft, automobiles, ball bearings, diesel motors, electrical equipment, explosives, fertilizers, furniture, glass, matches, paper and cardboard, plastics, plywood, precision tools, prefabricated houses, ships, steel, steelware, telephones, textiles, woodpulp; mining—copper, gold, iron ore, lead, uranium, zinc
Money: Basic unit—krona; one hundred öre equal one krona
Climates: Subarctic, continental moist

Sweden is a prosperous industrial nation. Its way of life has been called the "middle way" because it combines private enterprise with a government that greatly influences the economy. A high standard of living has spread to all income groups by means of a huge government welfare system that provides free education, largely free medical services, and pensions to the elderly, widows, orphans, and the like.

About 98 per cent of Swedes are Lutherans. The people are an athletic group and like outdoor activities. Cross-country skiing and hockey are the chief winter sports. Camping, hiking, gymnastics, swimming, soccer, and yachting are also popular.

Switzerland

Switzerland (142) is located in
central Europe. It is bordered by
West Germany, Austria, Liechtenstein,
Italy, and France.

Capital: Bern
Official Names: In German, Schweiz;
in French, Suisse; in Italian, Svizzera
Official Languages: German, French, and Italian
Government: Federal republic—president serves
as head of state, but his duties are largely ceremonial; the Federal Council,
a 7-member Cabinet, serves in place of a single chief executive; a two-house
legislature is called the Federal Assembly; political powers are divided between the central government and regional governments
National Anthem: "Swiss Psalm"
Flag: A large white cross representing Christianity centered on a red field; flag is
square in shape rather than rectangular
Area: 15,941 sq. mi. (41,288 km^2), including 523 sq. mi. (1,355 km^2) of inland
water; greatest distances—east-west, 213 mi. (343 km); north-south, 138 mi.
(222 km)
Population: 1983 estimate—6,348,000; distribution—58 per cent urban, 42
per cent rural; density—399 persons per sq. mi. (154 per km^2)
Largest Cities: (1978 est.) Zurich (379,600), Basel (185,300), Geneva
(150,100), and Bern (145,500)
Economy: Mainly manufacturing and agriculture, with tourism and other services
also important
Chief Products: Agriculture—dairy products, fruits, hay, potatoes, wheat; manufacturing—chemicals, electrical equipment, industrial machinery, precision
instruments, processed foods, textiles, watches
Money: Basic unit—franc
Climates: Oceanic moist, highlands

Switzerland is a small European republic known for its beautiful, snow-
capped mountains and freedom-loving people. The country has long had
a policy of neutrality in the many wars that have raged in Europe. Switzerland provided safety for thousands who fled from the fighting or from
political persecution in World Wars I and II.

Neutrality helped the Swiss develop valuable banking services to people in countries throughout the world. Switzerland is a thriving industrial
nation. Its most well-known exports are watches, cheese, and chocolate.

The mountains and valleys of Switzerland provide a grand variety of
sports: skiing, bobsledding, camping, climbing, and hiking. Target shoot-
ing is extremely popular.

Syria

Syria (143) is located in south-
western Asia. It is bordered by
Turkey, Iraq, Jordan, Israel,
and Lebanon.

Capital: Damascus
Official Name: Al-Jumhuria Al-Ara-
bia Al-Suria (The Syrian Arab Republic)
Official Language: Arabic
Government: Republic—elected president serves
as head of state and is the country's most powerful government official; the
195-member People's Council makes the country's laws; the president
heads the Baath Party, which controls the nation's politics and its armed
forces, the party's major power base
National Anthem: "Homat El Diyar" ("Guardians of the Homeland")
Flag: Three horizontal stripes of red, white, and black (top to bottom), traditional
Arab colors; two green stars appear on the white stripe
Area: 71,498 sq. mi. (185,180 km^2); greatest distances—east-west, 515 mi. (829
km); north-south, 465 mi. (748 km); coastline—94 mi. (151 km)
Population: 1983 estimate—9,869,000; distribution—50 per cent rural, 50 per
cent urban; density—137 persons per sq. mi. (53 per km^2)
Largest Cities: (1980 est.) Damascus (1,200,000), Aleppo (961,000), Homs
(338,000), and Latakia (234,000)
Economy: Based mainly upon agriculture, commerce, and industry
Chief Products: Agriculture—cotton, wheat, barley, fruits and vegetables, tobacco, sugar beets, livestock; manufacturing—textiles, petroleum products,
processed foods, cement, glass, soap; mining—oil, natural gas, phosphates,
asphalt, iron ore
Money: Basic unit—pound
Climates: Subtropical dry summer, steppe, desert

Syria is an Arab country of rolling plains, fertile river valleys, and barren
deserts. From ancient times, it has been a trade route linking Africa, Asia,
and Europe. Most Syrians today are Muslim Arabs, but the population
includes many ethnic and religious minorities. About half of all workers
are farmers who chiefly raise cotton and wheat.

Many Syrian villagers live much as their ancestors did centuries ago,
farming small plots and building houses of stone or sun-dried mud bricks.
Many wear traditional clothing such as billowy trousers and a large cloth
head covering. Bedouins live in tents and move about the countryside
grazing livestock. Women have traditionally had little freedom, but there
is a trend toward change. In the newer sections of the cities, life resembles that in the West. The people live in modern houses or apartments
and work in fields such as government and industry.

Taiwan

Taiwan (144) is an island nation composed of one major island and several small islands located off the coast of mainland China.

Capital: Taipei
Official Language: Chinese
Government: Republic—president, country's most powerful government official, appoints a prime minister and many other government officials; a 1,250-member National Assembly elects the president and amends the Constitution; five branches of government, each headed by a yuan or council, include executive, judicial, legislative, control (which watches over activities of government officials), and examination (which gives tests used to hire and promote government workers)
National Anthem: "Chung Hwa Min Kuo Kuo Ko" (The National Anthem of the Republic of China)
Flag: Red with a white sun on a blue canton in the upper left-hand corner
Area: 13,885 sq. mi. (35,961 km^2); including the Pescadores islands, but excluding Matsu and Quemoy; greatest distances—north-south, 235 mi. (378 km); east-west, 90 mi. (145 km); coastline—555 mi. (893 km)
Population: 1983 estimate—18,783,000; distribution—66 per cent urban, 34 per cent rural; density—1,352 persons per sq. mi. (522 per km^2)
Largest Cities: (1977 est.) Taipei (2,196,237) and Kaohsiung (1,172,977)
Economy: Based mainly upon manufacturing
Chief Products: Agriculture—asparagus, bananas, mushrooms, pineapples, rice, sugar cane, sweet potatoes, tea; fishing—sardines, tuna; forestry—bamboo, camphor, plywood; manufacturing—cement, electrical machinery, fertilizer, plastics, television sets, textiles
Money: Basic unit—New Taiwan dollar or yuan
Climates: Tropical wet and dry, tropical wet

Taiwan is a mountainous island in the South China Sea, about 90 miles (140 kilometers) from China. The Chinese Nationalist government moved to Taiwan after the Chinese Communists conquered the mainland in 1949.

Most Taiwanese are of Chinese ancestry. Most city people in Taiwan wear Western-style clothing. Farmers and others who work in the hot sun are identified by the cone-shaped straw hats they wear. A third of the people farm and, by Asian standards, live well.

The Taiwanese speak various Chinese dialects, but almost all the people also use Mandarin, the official dialect. The majority of Taiwanese are Buddhists or Taoists.

Tanzania

Tanzania (145) is located in eastern Africa. It is bordered by Kenya, Uganda, Mozambique, Malawi, Zambia, Rwanda, Burundi, and Zaire.

Capital: Dar es Salaam
Official Name: United Republic of Tanzania
Official Languages: English and Swahili
Government: Republic—elected president serves
    as head of state and is assisted by two vice-presidents; a Cabinet, headed by the president, governs the country; the 189-member elected legislature is called the National Assembly
National Anthem: "Mungo Ibariki Africa" ("God the Almighty Bless Africa")
Flag: The flag combines the old Tanganyika and Zanzibar flag colors in diagonal stripes; green, in the upper left, is for land; blue, in the lower right, is for the sea; black diagonal stripe from lower left to upper right is for the people; two narrow gold stripes on either side of the black are for mineral wealth
Area: 364,900 sq. mi. (945,087 km^2); greatest distances—north-south, 760 mi. (1,216 km); east-west, 740 mi. (1,184 km); coastline—550 mi. (1,424 km)
Population: 1983 estimate—20,629,000; distribution—62 per cent rural, 38 per cent urban; density—57 persons per sq. mi. (22 per km^2)
Largest Cities: (1978 census) Dar es Salaam (870,000), Mwanza (171,000), and Tanga (144,000)
Economy: Based mainly on agriculture, with some mining
Chief Products: Agriculture—cassava, citrus fruits, cloves, coconuts, coffee, copra, corn, cotton, fibers, hides and skins, nuts, oilseeds, rice, sisal, sugar cane, tea, tobacco; manufacturing and processing—clove oil, coconut oil, lime juice and oil; mining—diamonds, gold, lead, mica, salt, silver, tin
Money: Basic unit—shilling
Climates: Steppe, tropical wet and dry

In 1964, Tanganyika on the African mainland and several offshore islands known as Zanzibar became one nation officially called the United Republic of Tanzania. The country has beautiful scenery and much wildlife—buffaloes, elephants, giraffes, leopards, and lions. Big-game safaris attract many hunters. Africa's highest mountain, Kilimanjaro, rises in the north. Lake Victoria, Africa's largest lake, borders part of the north.

Most Tanzanians are Africans who are divided into about 120 ethnic groups. There are some Asians and Arabs who work as traders, shopkeepers, or skilled workers. Most Africans on the mainland follow tribal religions, but some are Christians and others are Muslims.

The country is chiefly agricultural. The most important food crops are cassava, corn, and beans. Important exports are cloves, coffee, cotton, diamonds, and sisal fibers.

Thailand

Thailand (146) is located in Southeast Asia. It is bordered by Laos,
Cambodia, Malaysia, and Burma.

Capital: Bangkok
Official Name: Muang Thai (Land of the Free)
Formerly Called: Siam
Official Language: Thai
Government: Constitutional monarchy—male or
   female monarch serves as chief of state in an
   advisory role; prime minister actually runs the government and selects a
   Cabinet to assist him called the Council of Ministers; the country's two-house
   legislature is called the National Assembly
National Anthem: "Pleng Chart" ("National Anthem of Thailand")
Flag: Five horizontal stripes of red, white, blue, white, and red; red represents
   the nation; white, purity; and blue, the monarchy
Area: 198,457 sq. mi. (514,000 km²); greatest distances—north-south, 1,100 mi.
   (1,770 km); east-west, 480 mi. (772 km); coastline—1,635 mi. (2,631 km)
Population: 1983 estimate—50,027,000; distribution—86 per cent rural, 14
   per cent urban; density—251 persons per sq. mi. (97 per km²)
Largest Cities: (1980 census) Bangkok (5,153,902), Chiang Mai (100,146),
   and Hat Yai (98,091)
Economy: Mainly agriculture and manufacturing, with some mining
Chief Products: Agriculture—rice, cassava, corn, cotton, rubber, sugar cane,
   tobacco; manufacturing—automobiles, cement, drugs, electronic equipment,
   food products, paper, plywood, textiles; forestry and fishing—teak, bamboo,
   rattan, anchovies, mackerel, shellfish; mining—tin, bauxite, iron ore, lead,
   manganese, natural gas, precious stones, tungsten
Money: Basic unit—baht
Climates: Tropical wet and dry, tropical wet

Thailand, or Siam, is the only country in Southeast Asia that has never
been ruled by a European country. Rivers are important for local transportation and many of the people live on or near the riverbanks. Some
boats are floating markets. Thailand is also a land of forest-covered
mountains with many wild animals.

Most Thai live in villages. The people are light-hearted and pleasure
loving. Few are rich, but most have comfortable homes and enough food
and clothing. The majority are Buddhists. Graceful temple dances are
performed by professional dancers and lords and ladies of the king's
court. Precious stones and gold embroidery decorate their costumes.

Favorite sports include soccer and Thai-style boxing, in which oppo-
nents fight with both their hands and feet. Another sport, fish fighting,
involves betting on fish that attack each other in jars.

Togo

Togo (147) is a small nation located in western Africa. It is bordered by Upper Volta, Benin, and Ghana.

Capital: Lomé
Official Name: République du Togo (Republic of Togo)
Official Language: French
Government: Presidential regime—an elected army officer serves as president; other government officials include both army officers and civilians; country's only legal political party is called The Rally of the Togolese People
Flag: Five alternating horizontal stripes, three green and two yellow, with a white star on a red square in the upper left corner; green symbolizes hope and agriculture; yellow, faith; white, purity; and red, charity and fidelity
Area: 21,622 sq. mi. (56,000 km^2); greatest distances—north-south, 365 mi. (587 km); east-west, 90 mi. (145 km); coastline—40 mi. (64 km)
Population: 1983 estimate—2,793,000; distribution—83 per cent rural, 17 per cent urban; density—129 persons per sq. mi. (50 per km^2)
Largest Cities: (1977 est.) Lomé (229,400), Sokodé (33,500), and Palimé (25,500)
Economy: Based largely upon agriculture, with mining also important
Chief Products: Agriculture—cacao, cassava, coffee, copra, cotton, palm kernels and oil, peanuts; mining—phosphates
Money: Basic unit—franc
Climates: Steppe, tropical wet and dry

Most of Togo's people are black Africans. The ways of life in Togo reflect the fact that several different groups of people have settled there. The groups are, however, similar in physical type, occupation, and religion. A majority of people live in rural areas, work on family-owned farms, and practice traditional African religions.

Many of the southern Togolese wear European-style clothes. Some work for the government and others have small businesses. Most of Togo's Christians, who are primarily Roman Catholics, live in the south. Most of Togo's Muslims live in the north. The northerners dwell in villages made up of adobe houses with cone-shaped thatched roofs. Most wear a white cotton smock. Only about two-fifths of the school-age children in Togo attend primary school and roughly 1 per cent attend secondary school.

Togo is mainly an agricultural country, but fishing and phosphate mining are also important. Bauxite, chromium, and iron deposits, though discovered, are yet undeveloped.

Tonga

Tonga (148) is an island nation located in the South Pacific Ocean. It is composed of about 150 islands located 400 miles (640 kilometers) west of Fiji.

Capital: Nukualofa
Official Name: Kingdom of Tonga
Official Language: Tongan
Government: Constitutional monarchy—king rules and appoints a premier and six additional Cabinet members to assist him; the Legislative Assembly consists of the Cabinet, seven nobles elected by Tonga's hereditary nobility, and seven commoners elected by the people; a privy council has the power to make laws when the Legislative Assembly is not in session
National Anthem: "'E 'Otua Mafimafi" ("O God Almighty")
Flag: A red field and a white canton; a red cross in the canton symbolizes the Christian faith of the Tongans
Area: 270 sq. mi. (699 km²); greatest distances—north-south, 270 mi. (432 km); east-west, 155 mi. (248 km); coastline—162 mi. (419 km)
Population: 1983 estimate—98,000; distribution—64 per cent rural, 36 per cent urban; density—363 persons per sq. mi. (140 per km²)
Largest City: (1977 est.) Nukualofa (18,200)
Economy: Based almost entirely upon agriculture
Chief Products: Bananas, copra, sweet potatoes, tapioca
Money: Basic unit—pa'anga; the pa'anga equals the Australian dollar
Climate: Tropical wet

The first people to settle in Tonga were Polynesians who probably came from Samoa. Tonga is the only remaining kingdom of Polynesia, having had a nobility and kings or queens for all its history. A protectorate of Great Britain from 1900, Tonga became independent in 1970.

Almost all the people of Tonga are Polynesian Methodists. The majority of Tongans live in small rural villages and raise crops. The people also fish for seafood such as shark and tuna. Most of the islands have no running water, and many have no electricity. The law requires that all Tongan children from 6 to 14 years old attend school. Rugby and football are favorite school sports.

Fertile soils and a warm climate have made agriculture the basis of the economy. The government owns all the land and each male 16 or over is entitled to a plot that he rents from the government.

Trinidad and Tobago

Trinidad and Tobago (149) is an island nation composed of two islands
located in the West Indies just
off the coast of Venezuela.

Capital: Port-of-Spain
Official Language: English
Government: Republic—president,
   elected by Parliament, serves as
   head of state; prime minister, the leader of
   the majority party in Parliament, heads the
   government and appoints a Cabinet to assist him; two-house Parliament
   consists of a Senate whose members are appointed by leading government
   officials and a House of Representatives whose members are elected by the
   people
National Anthem: "Forged from the Love of Liberty"
Flag: A black stripe, bordered by white stripes, runs across a red field from the
   upper left to the lower right corner
Total Land Area: 1,980 sq. mi. (5,128 km^2); greatest distances—north-south,
   90 mi. (144 km); east-west 90 mi. (144 km); coastline—292 mi. (470 km)
Population: 1983 estimate—1,232,000; distribution—51 per cent rural, 49 per
   cent urban; density—622 persons per sq. mi. (240 per km^2)
Largest Cities: (1977 est.) Port-of-Spain (42,950) and San Fernando (36,650);
   (1970 census) San Juan (30,802)
Economy: Mainly oil production and refining
Chief Products: Asphalt, cocoa, oil, sugar
Money: Basic unit—West Indies dollar
Climate: Tropical wet and dry

Trinidad and Tobago is a republic that was formerly a colony of Great
Britain. Its people are descendants of two main groups: black slaves
brought from Africa to work on the plantations and Indians brought there
to fill a labor shortage. The country became independent in 1962. Its
economy is based on oil production and refining, asphalt, sugar, and
tourism.

During the 1970's black-power groups protested against widespread
unemployment and what they considered social and economic inequal-
ity. Racial tensions have eased but unemployment continues to be a major problem.

About 95 per cent of the people can read and write. Roman Catholics
make up the largest religious group, followed by Anglicans and Hindus.
Many people in the country play native musical instruments called *pans,*
which are made from empty oil drums. Trinidad is the home of the limbo
dance and a form of folk music called calypso.

Tunisia

Tunisia (150) is in northern Africa.
It is bordered by Libya and Algeria.

Capital: Tunis
Official Name: Al-Jumhuriyah al-Tun-
usiyah (Republic of Tunisia)
Official Language: Arabic
Government: Republic—elected president heads the government assisted
by an appointed Cabinet; president has exten-
sive authority and runs both the government
and the Socialist Destour party, the country's only effective political party;
the party carries out the president's policies at all levels of government; one-
house, elected legislature is called the National Assembly
Flag: A large white circle on a red field; a red crescent and star are inside the
circle; the red color and the crescent and star design come from the flags of
the Ottoman Turks; the crescent and star are emblems of the Muslim religion
Area: 63,170 sq. mi. (163,610 km^2); greatest distances—north-south, 485 mi.
(781 km); east-west, 235 mi. (378 km); coastline—639 mi. (1,028 km)
Population: 1983 estimate—7,083,000; distribution—52 per cent urban, 48 per
cent rural; density—111 persons per sq. mi. (46 per km^2)
Largest Cities: (1975 census) Tunis (550,404), Sfax (171,297), Sousse
(69,530), and Bizerte (62,856)
Economy: Based mainly upon agriculture, with very limited mining
Chief Products: Agriculture—barley, citrus fruit, olives, wheat, wine; mining—
iron, lead, lignite, phosphates, zinc; forestry—oak, pine
Money: Basic unit—dinar
Climates: Subtropical dry summer, steppe, desert

Formerly a colony of France, Tunisia became independent in 1956. The
new government introduced many social and economic reforms. It gave
voting rights to women and set up a national school system.

Tunisia is part of the Arab world. Small groups of Europeans, Jews,
and Berbers live there, but almost all Tunisians are Arabs and Muslims.
The official language is Arabic. About half the people live on farms and
in small towns. Their homes are stone houses, mud huts, and tents.
These people wear traditional Arab clothing: a long, loose gown and a
turban or skullcap. The rest of the people live in larger towns and cities.
Many wear European-style clothes.

Although Tunisia is mainly an agricultural country, it has a more bal-
anced economy than many of its neighbors. There is neither a large
wealthy class nor a large poor class, and land ownership is not concentrated in the hands of a small minority.

Turkey

Turkey (151) is located in south-western Asia and southeastern Europe. It is bordered by Russia, Iran, Iraq, Syria, Greece, and Bulgaria.

Capital: Ankara
Official Name: Türkiye Cumhuriyeti
  (Republic of Turkey)
Official Language: Turkish
Government: Republic (military rule)—Constitution
  calls for a bicameral parliament made up of a 450-member National Assembly and a 150-member Senate; majority party leader of Assembly serves as prime minister and can be removed by a vote of no-confidence; military coup in 1980 resulted in appointment of new cabinet and suspension of Constitution until new one can be drafted
National Anthem: "İstiklâl Marşi" ("Independence March")
Flag: A white crescent beside a white star on a field of red; the crescent and five-pointed star are symbols of the Islamic religion
Area: 301,382 sq. mi. (780,576 km^2); greatest distances—north-south, 465 mi. (748 km); east-west, 1,015 mi. (1,633 km); coastline—2,211 mi. (3,558 km)
Population: 1983 estimate—48,410,000; distribution—56 per cent rural, 44 per cent urban; density—161 persons per sq. mi. (62 per km^2)
Largest Cities: (1975 census) Istanbul (2,547,364) and Ankara (1,701,004)
Economy: Mainly agriculture, but with rapidly expanding industry
Chief Products: Agriculture—barley, corn, cotton, fruits, potatoes, sugar beets, wheat; manufacturing—fertilizers, iron and steel, machinery and metal products, motor vehicles, processed foods and beverages, pulp and paper products, textiles
Money: Basic unit—lira; one hundred kurus equal one lira
Climates: Subtropical moist, subtropical dry summer, steppe

About 90 per cent of the Turkish people are descendants of an Asian people called Turks. Nearly all the people are Muslims. Old customs and traditions are everywhere. Kurds form the largest minority group in Turkey. They are people who roam the countryside with their livestock.

A new republican government in 1920 set out to make Turkey a modern state. Women now have the right to vote, to divorce, and to receive alimony. The Kurds and other tribal people have been encouraged to come into the mainstream of Turkish life. Education has been stressed. Now about 65 per cent of the people can read and write.

Turkey's most important contribution to the arts is the large number of Turkish mosques, with their thin minarets. Richly colored ceramic tiles decorate many mosques and palaces. Turkish craftworkers make excellent dishes and other ceramics.

Tuvalu

Tuvalu (152) is a small country consisting of nine islands in the South Pacific Ocean. It lies about 2,000 miles (3,200 kilometers) northeast of Australia.

Capital: Funafuti
Formerly Called: Ellice Islands
Official Languages: Tuvaluan and English
Government: Constitutional monarchy—prime minister heads the government; 12-member, elected legislature chooses the prime minister; each island is administered by a 6-member council
Flag: A light blue field with nine scattered gold stars representing the nine islands of the group; the Union Jack appears in the upper left corner
Area: 10 sq. mi. (26 km²); greatest distances—north-south, 400 mi. (640 km); east-west, 350 mi. (560 km); coastline—9 mi. (24 km)
Population: 1983 estimate—7,000; density—697 persons per sq. mi. (269 per km²)
Largest City: (1973 census) Funafuti (871)
Economy: Based mainly on the export of copra and handicrafts
Chief Products: Bananas, coconuts, copra, hand-woven baskets and mats, taro
Money: Basic unit—tala
Climate: Tropical wet

Formerly ruled by Great Britain, Tuvalu became independent in 1978. Most of the people of Tuvalu are Polynesians. They live in villages, most of which cluster around a church and a meeting house. Tuvaluan houses have raised foundations, open sides, and thatched roofs. The main foods are bananas, coconuts, fish, and taro, a tropical plant with one or more edible, rootlike stems. The islanders raise pigs and chickens to eat at feasts. They usually wear light, bright-colored cotton clothing. All the islands but one have a government-sponsored elementary school. The people speak the Tuvaluan language, and many also know English. Both languages are used in official government business.

Most of the islands of Tuvalu are atolls, or ring-shaped coral reefs that surround lagoons. The country has a tropical climate. Coconut palm trees cover much of the country, and the islanders use the coconuts to produce copra, or dried coconut meat, which is their chief export.

Uganda

Uganda (153) is a small nation located in east-central Africa. It is bordered by Zaire, Sudan, Kenya, Tanzania, and Rwanda.

Capital: Kampala
Official Language: English
Government: Republic—headed by a president; 136-member National Assembly makes the country's laws; people elect 126 Assembly members, and the president appoints 10; the presidential candidate of the political party that wins the most seats in the Assembly becomes president
Flag: A crested crane in a white circle is centered on six alternating horizontal stripes of black, yellow, and red (top to bottom); black stands for Africa, yellow for sunshine, and red for brotherhood
Area: 91,134 sq. mi. (236,036 km^2); greatest distances—north-south, 390 mi. (624 km); east-west, 375 mi. (600 km)
Population: 1983 estimate—15,117,000; distribution—88 per cent rural, 12 per cent urban; density—164 persons per sq. mi. (64 per km^2)
Largest Cities: (1976 est.) Kampala (531,022) and Jinja (189,540)
Economy: Based mainly on agriculture, with very limited mining
Chief Products: Agriculture—bananas, cassava, coffee, cotton, sweet potatoes, tea, tobacco; mining—copper
Money: Basic unit—shilling
Climates: Steppe, tropical wet and dry

Uganda is a thickly populated country. Most Ugandans are black Africans. Nearly all of the more than 20 ethnic groups have their own language. The Ganda, also known as the Baganda, are the largest and wealthiest group. Most are prosperous farmers. Women do much of the work on the farms. Coffee, cotton, and tea are the chief cash crops, and food crops include bananas and vegetables. The Ganda live in houses that have mud walls and corrugated iron roofs.

Three other ethnic groups in southern Uganda are farmers like the Ganda. But several ethnic groups in the drier parts of the north lead wandering lives as herders. The majority of Ugandans practice traditional African religions. About a fourth are Christians and some are Muslims.

A former British protectorate, Uganda, won independence in 1962 and became a republic in 1967. In 1979, Idi Amin Dada, a military officer who had taken over Uganda and ruled as a dictator, was overthrown.

United Arab Emirates

The United Arab Emirates (154) is a federation of seven independent Arab states located in southwestern Asia. It is bordered by Qatar, Oman, and Saudi Arabia.

Capital: Abu Dhabi
Official Language: Arabic
Government: Federation— the seven
    emirs who govern each of the seven states form
    the Supreme Council of the United Arab Emirates;
    the council appoints a president to serve as the country's chief of state and
    a prime minister to head the government; a Cabinet assists the prime min-
    ister in supervising the federal ministries or departments; each emir appoints
    representatives to the federal legislature, called the Consultive Assembly
Flag: A vertical red stripe near the staff and three horizontal stripes of green,
    white, and black (top to bottom)
Area: 32,278 sq. mi. (83,600 km^2); greatest distances—north-south, 250 mi.
    (402 km); east-west, 350 mi. (563 km); coastline—483 mi. (777 km)
Population: 1983 estimate—1,256,000; distribution—77 per cent urban, 23 per
    cent rural; density—39 persons per sq. mi. (15 per km^2)
Largest City: (1975 est.) Abu Zaby (150,000)
Economy: Mainly oil production and refining
Chief Products: Agriculture—dates; fishing—fish, shrimp; mining—petroleum
Money: Basic unit—dirham; one hundred fils equal one dirham
Climate: Desert

Formerly a British protectorate, United Arab Emirates, or UAE, gained full independence in 1971. A ruler called an emir governs each of the seven states that make up the UAE. Each emir controls the state's internal affairs. The federal government controls the UAE's foreign affairs. Arabic is the official language. Most people are Muslims. Some wear Western clothing, but most prefer traditional Arab garments.

Before the mid-1900's, this region was one of the substantially underdeveloped areas in the world. Most people earned a living as farmers, fishermen, traders, or nomadic herders. The discovery of oil during the late 1950's brought sudden wealth to the region and led to the development of modern cities and towns. Thousands of people came from other countries to work in the oil industry. Apartments, schools, hospitals, and roads were built to meet the needs of the growing population.

United States

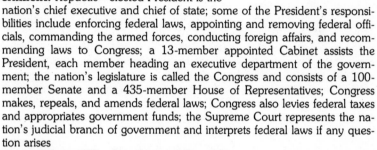

The United States (155) is located in North America. It is bordered by Canada and Mexico.

Capital: Washington, D.C.
Official Name: The United States of America
Also Called: The U.S.; the U.S.A.; America
Official Language: English
Government: Republic—elected President is the
nation's chief executive and chief of state; some of the President's responsibilities include enforcing federal laws, appointing and removing federal officials, commanding the armed forces, conducting foreign affairs, and recommending laws to Congress; a 13-member appointed Cabinet assists the President, each member heading an executive department of the government; the nation's legislature is called the Congress and consists of a 100-member Senate and a 435-member House of Representatives; Congress makes, repeals, and amends federal laws; Congress also levies federal taxes and appropriates government funds; the Supreme Court represents the nation's judicial branch of government and interprets federal laws if any question arises
National Anthem: "The Star-Spangled Banner"
Flag: Thirteen alternating horizontal stripes of red and white representing the thirteen original colonies; a blue canton with 50 white stars, one for each state of the Union
Area: 3,618,465 sq. mi. (9,371,781 km^2), including 74,389 sq. mi. (192,667 km^2) of inland water but excluding 60,788 sq. mi. (157,440 km^2) of Great Lakes and Lake Saint Clair and 13,942 sq. mi. (36,110 km^2) of coastal water; greatest distances excluding Alaska and Hawaii—east-west, 2,807 mi. (4,517 km); north-south, 1,598 mi. (2,572 km); greatest distances in Alaska—north-south, about 1,200 mi. (1,930 km); east-west, about 2,200 mi. (3,540 km); greatest distance in Hawaii—northwest-southeast, about 1,610 mi. (2,591 km); extreme points including Alaska and Hawaii—northernmost, Point Barrow, Alaska; southernmost, Ka Lae, Hawaii; easternmost, West Quoddy Head, Me.; westernmost, Cape Wrangell, Attu Island, Alaska; coastline—4,993 mi. (8,035 km), excluding Alaska and Hawaii; 12,383 mi. (19,929 km), including Alaska and Hawaii
Population: 1983 estimate—233,450,000; distribution—74 per cent urban, 26 per cent rural; density—65 persons per sq. mi. (25 per km^2)
Largest Cities: (1980 census) New York City (7,071,030), Chicago (3,005,072), Los Angeles (2,966,763), Philadelphia (1,688,210), and Houston (1,594,086)
Economy: Based mainly upon manufacturing, commerce, and trade

Chief Products: Agriculture—beef cattle, milk, corn, soybeans, hogs, wheat, cotton; fishing industry—shrimp, salmon, crabs; manufacturing—nonelectric machinery, transportation equipment, chemicals, food products, electric machinery and equipment, fabricated metal products, primary metals, printed materials, paper products, rubber and plastics products, clothing; mining—petroleum, natural gas, coal

Money: Basic unit—dollar

Climates: Extremely varied in the continental U.S.—continental moist, oceanic moist, subtropical moist, subtropical dry summer, steppe, desert, highlands, tropical wet and dry; Alaska—polar, subarctic; Hawaii—tropical wet

The United States is a large nation composed of 50 states and the District of Columbia. The District of Columbia is a segment of land set apart by the federal government as the nation's capital, called Washington, D.C. The country's federal system of government gives each state powers that in many other countries would be controlled by the national government. Each state, for instance, has broad control over public education and the establishment of civil and criminal laws.

The United States is the fourth largest country in the world both in area and population. Russia, Canada, and China have larger areas, and China, India, and Russia have more people. Covering the entire middle portion of North America, the United States stretches from the Atlantic Ocean in the east to the Pacific Ocean in the west. It also includes Alaska, in the continent's northwest corner, and Hawaii, a group of islands far out in the Pacific. Alaska and Hawaii are the two states most recently admitted to the Union, both in 1959.

A land of great beauty, the United States has a wide variety of sights and scenery. Included are huge, bustling cities; quaint, rural villages; vast, rolling fields of grain; and breathtaking, snow-capped mountains. The landscape ranges from the warm, sunny beaches of Florida, California, and Hawaii to the frozen northlands of Alaska; and from the colorful, dry desert areas of the Southwest to the rich forests of the Pacific Northwest and New England.

This huge country is extremely rich in the natural resources necessary for a productive economy. The economy of the United States is one of the most highly developed in the world and its people enjoy one of the world's highest standards of living. The country has great expanses of some of the most fertile soil on earth and its farms are the world's most productive. A plentiful water supply provides water for the nation's households, farms, and industries. The United States uses 400 billion gallons (1,500 billion liters) of water daily, 90 per cent of which is used to irrigate farms and operate busy factories. The factories of the United States, in fact, turn out the world's greatest abundance of goods. Large stretches of forests and rich fishing waters also contribute to the nation's productive economy. The largest quantities of fish are taken from the Gulf of Mexico. Underground lie valuable deposits of minerals such as coal, iron ore, natural gas, and petroleum, which are important for the nation's industrial strength. Even with its vast wealth of minerals, however, the United States still imports additional amounts of minerals in order to meet the country's needs.

The entire North American continent was largely a wilderness 350 years ago. Various tribes of Indians lived throughout the land between the Atlantic and Pacific oceans. Eskimos lived in what is now Alaska, and Polynesians inhabited the Hawaiian Islands. Europeans saw this land as a vast "new world" in which to build new and better lives. Thousands upon thousands of people from many countries came to the United States bringing ideas of freedom and equality.

During the 1500's, small groups of Spaniards settled in the southeastern and western portions of the country. It was not until the 1600's, however, that the English and other Europeans began colonizing the country's eastern coast. These pioneers faced great hardships and danger in this foreign wilderness including disease, lack of food, and Indian attacks. But with hard work and perseverance, the colonists soon established productive farms and plantations. They built towns, roads, churches, and schools; and they created many small industries. They also developed political practices and social beliefs that have had an important influence on the history of the United States. By the mid-1700's, most of these East Coast settlements had been formed into 13 British colonies, each with a governor and legislature but ultimately controlled by the British government. Britain began tightening its hold over the American Colonies, however, causing friction between the two sides. The Revolutionary War between the Americans and the British broke out on April 19, 1775. On July 4, 1776, the colonists boldly declared their independence and founded an independent nation based on freedom and equal opportunity for all. They finally defeated the British in 1783 and solidified their claim to independence.

Over the years, people from almost every part of the world continued to settle in the United States. They came seeking the golden opportunities that had become part of the American way of life. Due to this immigration, the United States has one of the world's most varied populations and has often been called "a nation of immigrants." Most white Americans trace their ancestry to Europe. At first, American settlers came from the countries of northern and western Europe including England, France, Germany, Ireland, The Netherlands, Scotland, and the Scandinavian lands of Denmark, Norway, and Sweden. In the late 1800's, large groups of southern and eastern Europeans began arriving, including some from Austria-Hungary, Greece, Italy, Poland, and Russia. A small percentage of Spanish-speaking Americans trace their ancestry back to Spain, but most are descendants of immigrants from Latin America. Most black Americans are descendants of Africans who were brought to the United States and forced into slavery during the 1600's, 1700's, and 1800's. After the 1800's, many Asians were attracted to the United States. Most came from China, Indochina, Japan, and the Philippines.

The people of the United States are of all races and backgrounds, but they have developed a common culture. Most have adopted the nation's customs, followed its traditions, and taken part in its politics. It is for this reason that the country has often been referred to as a "melting pot." Many Americans, however, still take special pride in their origins and preserve the traditions of their ancestors as well. Though the vast majority of Americans speak the same language, dress similarly, and eat the same kinds of foods, many have retained cultural features of their ancestors. In some cities, for instance, people of different national ethnic, or racial origins live in separate neighborhoods where shops and restaurants reflect their backgrounds. Ethnic festivals, parades, and other events across the nation emphasize this cultural diversity.

About three-fourths of the country's population are urban dwellers. This is a dramatic shift since the first census in 1790. At that time, only 5 per cent of Americans lived in cities. Several factors contributed to this population shift. First, agricultural methods and equipment improved greatly. Farm work became more efficient, production increased, and fewer farmers were needed. Second, an industrial boom at approximately the same time created many new jobs in the cities. Third, city life seemed glamorous to some rural people, particularly the young. Thus, many left home to seek excitement in the cities. Finally, as immigrants arrived in the United States, many found jobs in the cities and settled there.

Today, urban areas in the United States include large cities and their surrounding suburbs, as well as smaller, less congested towns. Networks of suburbs together with their central cities form units called metropolitan areas. There are about 320 metropolitan areas in the United States. There are jobs in these areas for a great variety of workers such as office and factory workers, bankers, doctors, fire fighters, police officers, medical personnel, teachers, trash collectors, and construction and transportation workers. Many Americans find urban areas exciting, interesting places to live. There are museums, art galleries, libraries, theaters, and concert halls that make these areas important cultural centers. There are generally a variety of specialized services available as well as a wide range of shops, restaurants, and recreation facilities.

About 98 per cent of all the United States land is classified as rural. Much of this land, however, is uninhabited. Roughly a fourth of all Americans live in rural areas, but only 9 per cent of them are farmers. Many others own or work in businesses related to agriculture such as grain and feed stores and warehouses. Others work in mining and related activities or in light industries. Still others work as teachers, police officers, sales-clerks, or other occupations. Some farmers hold additional jobs for part of the year to supplement their incomes. Farms, however, remain the economic basis of rural America.

American farmers today live very differently than their grandparents did. Modern machines, equipment, and methods have eliminated the backbreaking farm work that was a way of life many years ago. Machines now aid farmers in plowing, planting, harvesting, and delivering products to market. Conveyor systems help with feeding the animals, and milking machines make morning and evening chores easier. Most farm families enjoy all the comforts of people who live in cities—automobiles, tele-phones, radios, and televisions. In the 1900's, these conveniences brought rural families into close contact with the rest of the world.

Because of their small populations and small tax revenues, rural com-munities do not provide the variety of services offered in urban areas. Cultural and recreational facilities, therefore, are more limited. Social life, for many, centers around family gatherings, church and school activities, special interest clubs, and events such as state and county fairs. Rural areas also have economic divisions of wealthy, middle class, and poor, though the gaps are not so large as in urban areas. Most rural Americans live in comfortable, single-family houses. However, some people who live in scattered areas of the Appalachian Mountains and in other pockets of rural poverty have run-down houses and enjoy few luxuries.

Education is very important to most people of the United States, and Americans are among the best-educated people in the world. Almost all of the nation's children complete elementary school. About 75 per cent of them graduate from high school and about 60 per cent of high school graduates go on to colleges or universities. Approximately 15 per cent of the country's people finish at least four years of higher education. Not only schools, but libraries, museums, and various other educational insti-tutions provide learning opportunities for Americans of all ages. Museums and libraries throughout the nation offer classes, lectures, films, and field trips to interested citizens. The country has about 79,000 elementary schools, roughly 29,000 high schools, and nearly 3,000 colleges, univer-sities, and community colleges. Originally, most schools in the United States were privately owned and operated—many by church groups. The idea of free public education gained broad support in the early 1800's and soon state and local governments took the responsibility for setting up public school systems and establishing attendance laws. Today, 8 out of 10 of the nation's elementary and high schools, and about half of its colleges and universities are public institutions. The rest are private schools operated by religious groups or private organizations.

Adult education is also an important part of the nation's school system. Many adults attend classes at various schools, universities, recreation centers, or other institutions in order to improve their job skills, develop new hobbies, or learn more about interesting topics. Many men and women who previously held jobs or raised families return to school part time or full time to earn a degree.

Religion plays an important role in the lives of many Americans. It was also a major influence in the country's early history. Many American colonists had come to the New World to escape religious persecution. The U.S. Constitution, adopted almost 200 years ago, guarantees freedom of religion and provides that no religious group be given recognition as an official state church. Over the years, immigrants continued to flock to the United States because of this religious freedom. About three-fifths of the American population belong to organized religious groups. Most are Protestants and Roman Catholics. About 4 per cent are Jews and nearly 3 per cent belong to Eastern Orthodox Churches. In addition, small numbers practice Islam and Buddhism.

For recreation, Americans engage in a wide variety of activities. Sports, both spectator and competitive, rank as a leading pastime. Many Americans enjoy watching events such as automobile and horse races, and baseball, basketball, and football games—either in person or on television. Millions participate in bicycling, boating, bowling, fishing, golf, hiking, running, skiing, softball, swimming, and tennis as well as baseball, basketball, football, and soccer. Cultural events such as motion pictures, plays, concerts, operas, and dance performances are also popular. Hobbies such as gardening, coin collecting, stamp collecting, and photography occupy many Americans' leisure time. Interest in craft activities such as needlepoint, quilting, weaving, pottery making, and woodworking has increased in the past several decades. Many people travel in their leisure time, taking annual vacations as well as one-day or weekend trips. Camping is a popular way to travel among some Americans, while others prefer the conveniences of a hotel or motel.

The arts and sciences also have flourished in this wealthy land. The United States produced the electric light bulb, nuclear energy, the telephone, and the Salk polio vaccine. Americans developed the mass production system of manufacturing and were the first to set foot on the moon. Americans also developed unique art forms including jazz, musical comedy, and the skyscraper, which revolutionized urban architecture the world over.

Upper Volta

Upper Volta (156) is located in western Africa. It is bordered by Mali, Niger, Benin, Togo, Ghana, and the Ivory Coast.

Capital: Ouagadougou
Official Name: République de Haute-Volta (Republic of Upper Volta)
Official Language: French
Government: Republic (military rule)—president, a military officer, is head of state, head of the government, and head of the 17-member cabinet called the Council of Ministers; many of the ministers are also military officers
Flag: Three horizontal stripes of black, white, and red (top to bottom) stand for the Black, White, and Red Volta rivers
Area: 105,869 sq. mi. (274,200 km^2); greatest distances—east-west, 525 mi. (845 km); north-south, 400 mi. (644 km)
Population: 1983 estimate—7,439,000; distribution—91 per cent rural, 9 per cent urban; density—70 persons per sq. mi. (27 per km^2)
Largest Cities: (1977 est.) Ouagadougou (180,000), Bobo Dioulasso (120,000), and Koudougou (38,000)
Economy: Mainly farming and raising livestock
Chief Products: Agriculture—corn, cotton, fonio, livestock, millet, peanuts, rice
Money: Basic unit—franc
Climates: Steppe, tropical wet and dry

Landlocked Upper Volta is one of Africa's poor and underdeveloped countries. Because the country lacks rich soil and mineral deposits, its people have only the bare necessities of life.

Most of the population are black Africans who make their living by farming and raising cattle. For 800 years, the Mossi group have had a kingdom with a central government headed by the Moro Naba, their chief. The typical family lives in a *yiri,* a group of mud brick huts surrounding a small court, where sheep and goats are kept. The Bobo live in large villages where they build castlelike houses with clay brick walls and straw roofs. The Lobi have long been good hunters and farmers, but now they work as migrant laborers in and around the cities. There are also wandering Fulani shepherds, a few Hausa merchants, and the nomadic Tuareg herders. While there are Muslims and Christians in Upper Volta, most of the people practice traditional African religions.

Uruguay

Uruguay (157) is a small country
located on the southeastern coast
of South America. It is bordered by
Brazil and Argentina.

Capital: Montevideo
Official Name: La República Oriental
del Uruguay (The Eastern Republic
of Uruguay)
Official Language: Spanish
Government: Republic (military rule)—in 1973
military leaders took over the government and created a National Security
Council to manage the government; although there remains a president and
a Cabinet, the council rules by decree; council consists of the president,
certain Cabinet ministers, and representatives of the armed forces
National Anthem: "Himno Nacional del Uruguay" ("National Hymn of Uruguay")
Flag: A gold sun on a white canton appears on a field of nine alternating white
and blue horizontal stripes
Area: 68,037 sq. mi. (176,215 km^2); greatest distances—north-south, about 330
mi. (531 km); east-west, about 280 mi. (451 km); coastline—about 600 mi.
(966 km)
Population: 1983 estimate—2,952,000; distribution—84 per cent urban, 16
per cent rural; density—44 persons per sq. mi. (17 per km^2)
Largest Cities: (1975 census) Montevideo (1,229,750) and Salto (71,880)
Economy: Based mainly upon agriculture, with some manufacturing
Chief Products: Agriculture—linseed, meat, wheat, wool; manufacturing and
processing—canned meat, glass, leather, linseed oil, textiles; mining—gran-
ite, gravel, limestone, marble, sand
Money: Basic unit—peso
Climate: Subtropical moist

Uruguay has one of the highest living standards in South America. The
government operates free schools from kindergarten through college.
Most of the people speak Spanish and have Spanish or Italian ancestry.

The majority of Uruguayans live in clean and modern cities. Many people make their livings in commerce, manufacturing, government, construction, and mining.

Agriculture is the chief industry, and the vast pasturelands for livestock
are the chief wealth. Country life centers around farms, ranches, and
villages. Sheep and cattle graze throughout the year.

Most Uruguayans wear clothing similar to that worn in the United
States and Canada. They celebrate their most colorful festival, the *car-
naval,* during the three days before Lent.

Vanuatu

Vanuatu (158) is a small nation composed of 80 islands and located in the southwest Pacific Ocean.

Capital: Port-Vila
Formerly Called: New Hebrides
Official Languages: English and French
Government: Republic—president's
    role is largely ceremonial; a prime
    minister, the head of the majority party in
    Parliament, runs the government assisted by a Council of
    Ministers; the 39-member, elected Parliament makes the country's laws
National Anthem: "Yumi, yumi, yumi" ("We, we, we")
Flag: Two horizontal stripes of red and green; a black triangle near the mast extending into a black horizontal stripe between the red and green; a yellow band lies just inside the triangle and runs down the middle of the black stripe; a yellow spiral with two crossed feathers appears in the triangle
Total Land Area: 5,700 sq. mi. (14,763 km^2); greatest distances—north-south, 500 mi. (800 km); east-west, 250 mi. (400 km); coastline—976 mi. (1,562 km)
Population: 1983 estimate—125,000; distribution—77 per cent rural, 23 per cent urban; density—21 persons per sq. mi. (8 per km^2)
Largest City: (1979 census) Vila (10,158)
Economy: Based mainly upon agriculture, with tourism also important
Chief Products: Agriculture—copra, cocoa, coffee, yams, taro, manioc, sweet potatoes, breadfruit, cattle; manufacturing and industry—processed fish and meat, bricks, cement, lumber, tourism; mining—manganese
Money: Basic unit—Vanuatu franc (one franc equal one hundred centimes)
Climate: Tropical wet

The mountainous islands of Vanuatu form a Y-shaped chain extending about 500 miles (800 kilometers) in a north-south direction. On several of the islands, active volcanoes rise among the mountains above narrow coastal plains. The country's climate is tropical and both tourists and islanders enjoy the surrounding sparkling blue waters and sandy beaches.

The majority of Vanuatu's people are Melanesians. Asians, Europeans, and Polynesians make up about 10 per cent of the population. Most are Christians. A few practice local religions. Over 100 languages are spoken throughout the country, but Bislama, a combination of English words and Melanesian grammar, is the language most commonly used.

About three-fourths of the people of Vanuatu live in rural villages. Their houses are made of wood from nearby forests and of bamboo and palm leaves. They produce nearly all the food they need, growing fruits and vegetables, raising chickens and hogs, and catching fish.

Vatican City

Vatican City (159) lies entirely
within the city of Rome, Italy. It
is under the direction of the pope.

Official Name: Stato della Città
del Vaticano (The State of Vatican City)
Also Called: The Vatican
Official Language: none
Administration: Pope is absolute ruler and
heads all branches of government; other of-
ficials have delegated authority; a Governor directs internal domestic affairs
much as a city mayor would; Cardinal Secretary of State handles foreign
affairs and coordinates ecclesiastical and political affairs
Flag: The pope's yellow and white square banner is the official state flag; left
half is yellow and the right half is white; emblem with crossed keys, representing Saint Peter, and a tiara, representing the pope's superiority, appear
in the white portion
Population: 1983 estimate—1,000; density—5,884 persons per sq. mi. (2,272
per km^2)
Interesting Sights: The Sistine Chapel in the Vatican Palace noted for paintings
by Michelangelo that decorate the ceiling and rear wall; St. Peter's Church
and St. Peter's Square
Economy: All residents connected with activities of the Roman Catholic Church
Money: Basic unit—none
Climate: Subtropical dry summer

Vatican City is the smallest independent state in the world. It was created
with the signing of the Treaty of Lateran in 1929. It serves as the spiritual
and governmental center of the Roman Catholic Church, the largest
Christian denomination.

The ruler of Vatican City is the pope, who delegates most of his tem-
poral authority to other officials, devoting his time primarily to spiritual
and ecclesiastical matters. Vatican City has a Governor, a Cardinal Secretary of State, and civil law courts, as well as the Tribunal of the Sacred
Roman Rota, which handles religious cases.

High stone walls surround most of Vatican City. Huge St. Peter's
Church, with its stately dome, dominates the complex. It is the largest
Christian church in the world. Vatican Palace is a group of connected
buildings with chapels, apartments, and museums—all clustered around
open courts. Vatican museums have priceless collections of statuary and
paintings. Vatican Archive and Vatican Library contain important religious and historical documents, manuscripts, and books.

172

Venezuela

Venezuela (160) is on the northern coast of South America. It is bordered by Colombia, Brazil, and Guyana.

Capital: Caracas
Official Name: República de Venezuela (Republic of Venezuela)
Official Language: Spanish
Government: Federal republic—elected president serves as head of state and as head of the executive branch of government; the National Congress is the country's two-house legislature and consists of the Chamber of Deputies and the Senate
National Anthem: "Gloria al Bravo Pueblo" ("Glory to the Brave People")
Flag: Three horizontal stripes of yellow, blue, and red (top to bottom); seven white stars form an arch on the blue stripe; the country's coat of arms appears in the top left corner of the flag
Area: 352,145 sq. mi. (912,050 km^2); greatest distances—north-south, 790 mi. (1,271 km); east-west, 925 mi. (1,489 km); coastline—1,750 mi. (2,816 km)
Population: 1983 estimate—15,203,000; distribution—76 per cent urban, 24 per cent rural; density—44 persons per sq. mi. (17 per km^2)
Largest Cities: (1978 est.) Caracas (1,279,600), Maracaibo (845,000), Valencia (471,000), and Barquisimeto (459,000)
Interesting Sights: The Sabana Grande, one of the main shopping and business districts of Caracas; spectacular modern architecture on the campus of the Central University of Venezuela in Caracas
Economy: Mainly agriculture and oil production
Chief Products: Agriculture—cotton, sugar cane, corn, coffee, rice; manufacturing—refined petroleum products, petrochemicals, processed foods, textiles; mining—petroleum, iron ore, diamonds, gold
Money: Basic unit—bolívar
Climates: Steppe, tropical wet and dry, tropical wet

Venezuela was one of the poor countries in South America until the 1920's, when its petroleum industry began to boom. Since then, the country has become one of the wealthy nations on the continent.

Before the 1500's, numerous Indian tribes lived in what is now Venezuela. When the Spanish colonized the area, they conquered the tribes and imported black slaves from Africa. Many of the Indians, Spaniards, and blacks intermarried. Today, about 65 per cent of the people are of mixed ancestry. Roman Catholicism is the traditional religion.

Compared with some other Latin-American countries, Venezuela is not rigidly segregated on the basis of racial or class differences. Poverty, however, remains a major problem.

Vietnam

Vietnam (161) is in the eastern part of the Indochinese Peninsula in Southeast Asia. It is bordered by China, Laos, and Cambodia.

Capital: Hanoi
Official Name: Cong Hoa Xa Hoi Chu Nghia Viet Nam (Socialist Republic of Vietnam)
Official Language: Vietnamese
Government: Communist dictatorship—leaders of the nation's Communist Party tightly control the government; the party's 15-member Politburo is the most powerful governmental unit; the 496-member National Assembly meets to endorse laws and policies made by the party; a Council of State, consisting of members of the National Assembly, serves as a collective presidency and deals with matters such as national defense and execution of laws; members of a Council of Ministers head government departments
Flag: A large yellow star representing Communism lies centered on a red field
Area: 127,242 sq. mi. (329,556 km²); greatest distances—north-south, 1,030 mi. (1,657 km); east-west, 380 mi. (612 km); coastline—1,435 mi. (2,309 km)
Population: 1983 estimate—57,990,000; distribution—77 per cent rural, 23 per cent urban; density—456 persons per sq. mi. (176 per km²)
Largest Cities: (1976 census) Ho Chi Minh City (3,460,500), Hanoi (1,443,500), and Haiphong (1,190,900)
Economy: Based mainly upon agriculture
Chief Products: Agriculture—rice; manufacturing—cement, iron and steel, paper, textiles; mining—coal
Money: Basic unit—dong
Climates: Subtropical moist, tropical wet and dry, tropical wet

Vietnam has in recent times been through a devastating civil war. United States troops joined South Vietnam in the fighting against Communist North Vietnam. In 1973, South Vietnam, the United States, and the Communists signed a cease-fire agreement, and the United States removed its last combat troops. The Communists soon launched another offensive against South Vietnam, and in 1975 gained control of the entire country. Vietnam is now a Communist dictatorship.

Two-thirds of the Vietnamese are farmers. Most of those who practice a religion are Buddhists. Many also worship the spirits of animals and plants and believe in the teachings of Confucianism and Taoism. The government, however, discourages religion.

In spite of a shortage of trained teachers, most Vietnamese can read and write. Their chief foods are fish, rice, and vegetables.

Western Samoa

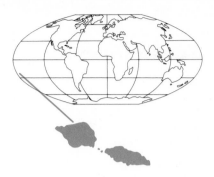

Western Samoa (162) is an island nation in the South Pacific Ocean. It lies about 1,700 miles (2,740 kilometers) northeast of New Zealand. It is composed of two main islands and several smaller islands.

Capital: Apia
Official Languages: Samoan and English
Government: Parliamentary—head of state presently holds office for life; when he dies, the 47-member Legislative Assembly will elect a head of state every five years from one of the country's two royal families; prime minister, assisted by a Cabinet, runs the government
Flag: A red field with a blue canton in the upper left-hand corner; five white stars on the canton symbolize the Southern Cross constellation
Area: 1,097 sq. mi. (2,842 km^2); greatest distances—east-west, on each of the two main islands, 47 mi. (76 km); north-south, 15 mi. (24 km) on Upolu, 27 mi. (43 km) on Savai'i; coastline (total for both islands)—about 230 mi. (370 km)
Population: 1983 estimate—160,000; distribution—78 per cent rural, 22 per cent urban; density—145 persons per sq. mi. (56 per km^2)
Largest City: (1976 census) Apia (32,099)
Economy: Based mainly on agriculture
Chief Products: Agriculture—bananas, cacao, coconuts
Money: Basic unit—tala
Climate: Tropical wet and dry

Samoans are tall, brown-skinned Polynesians. Most can read and write Samoan. About half can read and write English.

Samoans live simply, with life centering around the family. Their houses are open-sided with thatched roofs supported by poles. Most Samoan men and some women wear only a *lava-lava,* a piece of cloth wrapped around the waist like a skirt. Most of the women wear dresses or a skirt and blouse.

Samoans greatly enjoy dancing. They also love to play their own version of cricket, a game they learned from English missionaries. All Samoans are Christians.

By some standards, Western Samoa is a poor, developing country. The people have little cash income, but most have little need for money. They live by raising their own food on small plots. They build their own houses and make most of their own clothing.

Yemen (Aden)

Yemen (Aden) (163) is in the southern part of the Arabian Peninsula in southwest Asia. It is bordered by Saudi Arabia, Oman, and Yemen (Sana).

Capital: Aden
Official Name: Jumhuriyyat al-Yemen ash-Shaabiyyah al-dimugratiyyah (People's Democratic Republic of Yemen)
Also Called: South Yemen, Southern Yemen
Official Language: Arabic
Government: Republic—the United Political Organization National Front (UPONF), the country's only political party, governs; a three-member Presidential Council heads the government; the council's chairman appoints a Cabinet to help carry out government operations
Flag: Three horizontal stripes of red, white, and black (top to bottom); a small red star lies centered on a light blue triangle near the mast
Area: 128,560 sq. mi. (332,968 km^2); greatest distances—north-south, 435 mi. (696 km); east-west, 680 mi. (1,088 km); coastline—about 740 mi. (1,191 km)
Population: 1983 estimate—2,083,000; distribution—63 per cent rural, 37 per cent urban; density—16 persons per sq. mi. (6 per km^2)
Largest City: (1977 est.) Aden (271,600)
Economy: Largely undeveloped except for oil refining and port facilities
Chief Products: Agriculture—barley, cotton, dates, millet, sorghum, wheat; industry—dyeing, fishing, oil refining, ship refueling, tanning, weaving
Money: Basic unit—dinar
Climates: Steppe, desert

Yemen (Aden) is hot and dry. About 90 per cent of the people are Arabs. Their religion is Islam. Approximately 10 per cent of the people can read and write.

Aden is the country's capital and largest city. Its oil refinery and port provide Yemen (Aden) with most of its income. Some of the people in this city wear Western-style clothing, live in modern houses or apartments along broad streets, and shop in supermarkets. Others follow an older way of life. They live in thick-walled houses along narrow, twisting alleys, and they shop in open-air markets. Most of the men wear the striped *futa*, or "kilt." Women appear in veils and dark, shapeless clothing.

On the coast, the people live by fishing. Farmers live inland in the valleys and on a few scattered oases. Herders in the desert search constantly for water and food for their sheep and goats.

Yemen (Sana)

Yemen (Sana) (164) is located on the southwestern edge of the Arabian Peninsula in southwest Asia. It is bordered by Saudi Arabia and Yemen (Aden).

Capital: Sana
Official Name: Al-Jumhuriyah al Arabiyah al Yamaniyah (The Yemen Arab Republic)
Also Called: North Yemen; Northern Yemen
Official Language: Arabic
Government: Military rule—in 1974, army leaders suspended the nation's Constitution and dissolved the country's legislature; a council composed of military leaders controls the government; the head of the council serves as the country's president
National Anthem: "Assalam Alwatani Al-Gumhuri" ("Republican National Anthem")
Flag: Three horizontal stripes of red, white, and black (top to bottom); a green star appears in the center of the white stripe
Area: 75,290 sq. mi. (195,000 km^2); greatest distances—north-south, 350 mi. (560 km); east-west, 260 mi. (416 km); coastline—about 280 mi. (451 km)
Population: 1983 estimate—6,270,000; distribution—90 per cent rural, 10 per cent urban; density—83 persons per sq. mi. (32 per km^2)
Largest Cities: (1979 est.) Sana (192,045); (1978 est.) Hodeida (106,080)
Economy: Based mainly upon agriculture and very limited light industry
Chief Products: Agriculture—coffee, fruits, grains, qat, vegetables; manufacturing—handicrafts
Money: Basic unit—rial
Climates: Steppe, desert

Most of the people of Yemen (Sana) are Arabs. The men wear cotton breeches and shirts. The women wear long robes, black shawls, and veils. In the High Yemen, people build mud or stone houses. In the Tihamah, to the south, most people live in straw huts. People in the cities often build one-story mud brick houses. Wealthy Yemenis live in houses that are two to six floors high. They wear white silk robes, turbans, and leather sandals. Most people are Muslims. They are divided into two different sects, and their differences have caused much bitterness in the country.

Only about 10 per cent of the Yemenis can read and write. Farmers are the largest group of workers. Yemen (Sana) has almost no industry. Most goods are made by hand. The people weave and dye cloth, make rope, glassware, harnesses, saddles, and pottery.

177

Yugoslavia

Yugoslavia (165) is located on the Balkan Peninsula in south-central Europe. It is bordered by Italy, Austria, Hungary, Romania, Bulgaria, Greece, and Albania.

Capital: Belgrade
Official Name: Socijalistička Federativna Republika Jugoslavija (Socialist Federal Republic of Yugoslavia)
Official Languages: Serbo-Croatian, Slovenian, Macedonian
Government: Socialist republic—headed by a nine-member council called the Presidency, the country's chief policymaking body; council members take turns serving as president of the council and as head of state, each for one year; a two-house legislature called the Federal Assembly passes the country's laws and elects a cabinet called the Federal Executive Council that administers government departments; the president of the Federal Executive Council is called the premier and heads the government
National Anthem: "Hej Sloveni" ("Hey Slavs")
Flag: Three horizontal stripes of blue, white, and red (top to bottom); a red star outlined in yellow symbolizes Communism and lies in the center of the flag
Area: 98,766 sq. mi. (255,804 km^2); greatest distances—north-south, 415 mi. (668 km); east-west, 475 mi. (764 km); coastline—490 mi. (789 km)
Population: 1983 estimate—23,129,000; distribution—58 per cent rural, 42 per cent urban; density—233 persons per sq. mi. (90 per km^2)
Largest Cities: (1971 census) Belgrade (746,105) and Zagreb (566,224)
Economy: Based mainly upon manufacturing and mining, with some agriculture
Chief Products: Agriculture—corn, livestock, potatoes, sugar beets, wheat; manufacturing—automobiles, chemicals, food products, machinery, metal products, textiles, wood products
Money: Basic unit—dinar
Climates: Continental moist, oceanic moist, steppe, highlands

About 85 per cent of Yugoslavia's people belong to six Slavic nationality groups: Serbs, Croats, Slovenes, Bosnian Muslims, Macedonians, and Montenegrins. Since 1945, the country has developed its own, independent form of Communism. Before the Yugoslav Communists gained control, most of the people were poor farmers. The Communists have encouraged industrial growth. They have developed an unusual system of economic self-management under which the workers themselves run the industries. The standard of living in Yugoslavia is fairly high. Most people belong to Eastern Orthodox churches. Much of Yugoslavia's art reflects the rich folk traditions of the country.

Zaire

Zaire (166) is located in south-central Africa. It is bordered by the Congo, the Central African Republic, Sudan, Uganda, Rwanda, Burundi, Tanzania, Angola, and Zambia.

Capital: Kinshasa
Formerly Called: Belgian Congo; Congo
Official Language: French
Government: Presidential regime—elected president has almost complete control and makes all major government policy decisions; a 240-member legislature, the National Legislative Council, meets to debate budget details and to pass laws proposed by the president; president appoints a commissioner to head each of the government's 20 executive departments
Flag: A field of green; a hand holding a flaming torch in a yellow circle appears in the center of the flag; the torch stands for progress and honors those who died in the nation's conflicts
Area: 905,568 sq. mi. (2,345,409 km^2); greatest distances—north-south, about 1,300 mi. (2,090 km); east-west, about 1,300 mi. (2,090 km); coastline—25 mi. (40 km)
Population: 1983 estimate—30,824,000; distribution—60 per cent rural, 40 per cent urban; density—34 persons per sq. mi. (13 per km^2)
Largest Cities: (1974 est.) Kinshasa (2,008,352), Kananga (601,239), and Lubumbashi (403,623)
Economy: Based mainly upon the mining industry, with very limited manufacturing
Chief Products: Agriculture and forestry—cassava, cocoa, coffee, cotton, corn, palm oil, rice, rubber, tea, timber; manufacturing—beer, cement, soap, soft drinks, steel, textiles, tires; mining—cadmium, cobalt, copper, gold, industrial diamonds, manganese, oil, silver, tin, zinc
Money: Basic unit—zaire
Climates: Tropical wet and dry, tropical wet

This large nation in the heart of Africa was called the Congo until 1971. Zaire's people are black Africans who are members of different ethnic groups with their own languages. The government has made progress toward giving the people a sense of national unity.

With hand tools, most rural Zairians farm small plots. The people live in mud brick houses with thatched roofs. In the cities, important government officials, business managers, and merchants live in attractive bungalows. About 60 per cent of the people practice traditional African religions based on the worship of many gods and spirits. Christians make up about 40 per cent of worshipers.

Zambia

Zambia (167) is in south-central Africa. It is bordered by Zaire, Tanzania, Malawi, Mozambique, Zimbabwe, Botswana, Namibia, and Angola.

Capital: Lusaka
Official Name: Republic of Zambia
Formerly Called: Northern Rhodesia
Official Language: English
Government: Republic—elected president serves as head of state and government; the 125-member legislature is called the National Assembly; president appoints a Cabinet from among the Assembly; country has only one political party, the United National Independence Party, whose leaders have a key role in establishing government policies
Flag: A field of green for natural resources; three vertical stripes of red for freedom, black for the people, and orange for mineral wealth are in the lower right corner; an orange eagle appears above the stripes
Area: 290,586 sq. mi. (752,614 km^2); greatest distances—east-west, 900 mi. (1,448 km); north-south, 700 mi. (1,127 km)
Population: 1983 estimate—6,243,000; distribution—60 per cent rural, 40 per cent urban; density—21 persons per sq. mi. (8 per km^2)
Largest Cities: (1980 est.) Lusaka (641,000), Kitwe (341,000), Ndola (323,000), and Chingola (192,000)
Economy: Mainly mining with important construction and agriculture
Chief Products: Agriculture—cassava, cattle, corn, cotton, millet, tobacco; fishing—perch, whitebait; manufacturing and processing—cement-making, copper-smelting, sawmilling; mining—cobalt, copper, lead, zinc
Money: Basic unit—kwacha
Climate: Tropical wet and dry

Landlocked Zambia ranks as one of the largest producers of copper. The busy copper mines help make Zambia a rich African country. Most Zambians are black Africans who speak Bantu languages. There are more than 70 ethnic groups represented, and eight major local languages are spoken in Zambia. Many people also speak English, the official language.

In remote parts of Zambia, village life goes on much as it has for hundreds of years. Most of the people live in the bush, as the rural areas are called. The homes are circular, grass-roofed huts. The people raise food crops on the surrounding land. Maize is the main food. A favorite dish is nshima, a thick porridge made from maize.

Most Zambians are Christians, but traditional local beliefs still have a strong hold on villagers. Witchcraft, however, and customs such as polygyny, or marrying several wives, are dying out slowly in the towns.

Zimbabwe (Rhodesia)

Zimbabwe (168) is located in southern Africa. It is bordered by Zambia, Mozambique, the Republic of South Africa, and Botswana.

Capital: Salisbury
Formerly Called: Rhodesia
Official Language: English
Government: Republic—a two-house
Parliament, consisting of the House of Assembly and the Senate, makes the country's laws; president, elected by the House, serves as head of state; the head of the political party with the most seats in the House serves as prime minister, the country's chief executive; prime minister and an appointed Cabinet carry out government operations
Flag: Seven horizontal stripes of green, yellow, red, black, red, yellow, and green; a white triangle on the left contains a yellow Great Zimbabwe bird on a red 5-pointed star
Area: 150,804 sq. mi. (390,580 km^2); greatest distances—north-south, 470 mi. (752 km); east-west, 515 mi. (824 km)
Population: 1983 estimate—7,926,000; distribution—80 per cent rural, 20 per cent urban; density—49 persons per sq. mi. (19 per km^2)
Largest Cities: (1979 est.) Salisbury (118,500) and Bulawayo (85,700)
Economy: Agriculture, mining, and manufacturing
Chief Products: Agriculture—cattle, coffee, corn, cotton, sugar, tea, tobacco, wheat; manufacturing and processing—chemicals, clothing and footwear, iron and steel, metal products, processed foods, textiles; mining—asbestos, chrome, coal, copper, gems, gold
Money: Basic unit—Zimbabwe dollar
Climates: Steppe, tropical wet and dry

Zimbabwe, formerly called Rhodesia, is a nation populated largely by black Africans. Whites, Asians, and persons of mixed ancestry form about 4 per cent of the population. Most of the blacks live in rural areas, where they grow only enough corn and other food crops to feed themselves. The minority groups live mainly in Zimbabwe's cities and towns. Some whites are farmers. English is the official language of Zimbabwe, but most of the people speak African languages.

Zimbabwe is a leading producer of gold, chrome, and asbestos. It also mines iron ore, coal, tin, copper, and precious stones. Corn is Zimbabwe's chief food crop. Farmers also grow sugar, tobacco, cotton, peanuts, tea, and wheat. Some raise cattle.

Statistics and maps

This useful section features aids in trying to learn about the nations of the world. Included is information about the seven continents into which the world is divided, as well as information about political units of the world other than the independent nations discussed in this volume. The nation locator maps pinpoint the position of each nation on its continent or its location in the Pacific Ocean.

The table on page 183 shows how continental statistics, or numerical facts, compare. For instance, although South America is almost 3 million square miles (7 million square kilometers) larger in area than Europe, it has approximately only one-third of Europe's population. Almost twice the size of North America, Asia has over four times as many people living in each square mile or square kilometer of its territory.

Also on page 183 is a table of other political units of the world. These colonies and territories, also called dependencies, are controlled in some way by independent countries. In most of them, an independent nation is responsible for internal affairs, foreign affairs, and defense. Some of these units have internal self-government, while their foreign affairs and defense are administered by an independent country. The table shows which independent nation holds that responsibility for each colony or territory. The table also shows the size of each colony or territory in area and population.

The nation locator maps begin on page 184 with a global map of the continents and a table of nations by continent. This table serves as a quick reference for which nations are located on each continent. Because the frozen continent of Antarctica has no recognized independent nations, it is not included in the table or in the individual continent maps that follow.

Comparing the makeup of the continents is interesting. Africa, for example, has the most nations of all, with over 50. The continent of Australia is the location of only one country, also called Australia. However, a number of small island nations are not located on any continent, but lie in the Pacific Ocean to the north and east of the island continent of Australia. For this reason, the Pacific Islands are listed with Australia and often called Oceania. Some island nations in the Pacific are not part of Oceania, such as Indonesia, Japan, and the Philippines. They are grouped with the continent of Asia.

Beginning on page 185 you will find numbered continental maps with lists of the nations located on those continents. Notice that Russia appears on two maps, those of Europe and Asia. The largest country in the world in area, Russia stretches over both these continents. The number following each nation in the lists corresponds to the country's position on the map. It also corresponds with the number shown in the individual nation entries that comprise the first part of this volume.

The map of Europe and some other maps illustrate part of another continent (part of Asia in the case of the European map), but only the independent countries located on the named continent are keyed. Some familiar land areas are missing, such as Greenland or other colonies and territories listed in the table on page 183. This is because, as discussed earlier, they are not independent nations of the world.

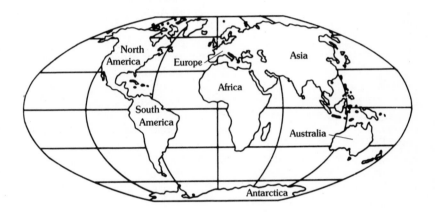

Continents of the world

Name	Area*		Population*	Density†	
	In sq. mi.	In km²		Per sq. mi.	Per km²
Africa	11,714,000	30,339,000	514,000,000	44	17
Antarctica	5,100,000	13,209,000	—	—	—
Asia	17,011,000	44,059,000	2,810,000,000	166	64
Australia	2,966,000	7,682,000	15,000,000	5	2
Europe	4,063,000	10,524,000	674,000,000	166	64
North America	9,400,000	24,345,000	387,000,000	41	16
South America	6,883,000	17,828,000	255,000,000	36	14

*Area figures are rounded to the nearest thousand; and population figures, to the nearest million.
†Density refers to the number of persons per unit of land area.

Other political units of the world*

Name	Area		Population†
	In sq. mi.	In km²	
American Samoa (U.S.)	76	197	34,000
Andaman and Nicobar Is. (India)	3,202	8,293	115,133

*The political units listed are administered by the country shown in parentheses.
†Populations are 1983 and earlier estimates based on the latest figures from official government and United Nations sources.

Other political units *(cont.)*

Name	Area		Population†
	In sq. mi.	In km²	
Anguilla (G.B.)	35	91	6,500
Azores (Port.)	905	2,344	273,400
Bermuda (G.B.)	21	54	63,000
British Indian Ocean Territory (G.B.)	23	60	2,000
Brunei (G.B.)	2,226	5,765	277,000
Canary Is. (Sp.)	2,808	7,273	1,487,000
Cayman Is. (G.B.)	100	259	19,000
Channel Is. (G.B.)	75	195	133,000
Cook Is. (N.Z.)	91	236	20,000
Easter I. (Chile)	63	163	2,600
Faeroe Is. (Denmark)	540	1,399	41,000
Falkland Is. (G.B.)	4,700	12,173	2,000
French Guiana	35,135	91,000	67,000
French Polynesia	1,544	4,000	142,000
Gaza Strip (Egypt)‡	146	378	467,000
Gibraltar (G.B.)	2.3	6	35,000
Greenland (Den.)	840,004	2,175,600	52,000
Guadeloupe (Fr.)	687	1,779	373,000
Guam (U.S.)	212	549	113,000
Hong Kong (G.B.)	1,126	2,916	5,315,000
Macao (Port.)	6	16	286,000
Madeira Is. (Port.)	308	797	245,000
Man, Isle of (G.B.)	227	588	73,000
Martinique (Fr.)	425	1,102	381,000
Midway I. (U.S.)	2	5	2,220
Montserrat (G.B.)	38	98	12,000
Namibia, or South West Africa (S. Af.)	318,261	824,292	1,097,000
Netherlands Antilles	383	993	256,000
New Caledonia (Fr.)	7,376	19,103	148,000
Niue I. (N.Z.)	100	259	4,000
Norfolk Is. (Austral.)	14	36	2,000
Pacific Is., Trust Territory of the (U.S.)	717	1,857	119,440
Caroline Is.	463	1,199	75,394
Mariana Is.	184	477	14,335
Marshall Is.	70	181	29,511
Pitcairn Islands Group (G.B.)	2	5	63
Puerto Rico (U.S.)	3,435	8,897	3,188,000
Reunion (Fr.)	969	2,510	539,000
St. Christopher-Nevis (G.B.)	101	262	41,000
Saint Helena (G.B.)	162	419	6,300
Saint Pierre and Miquelon (Fr.)	93	242	5,000
South West Africa (S. Af.), see Namibia			
Tokelau (N.Z.)	4	10	2,000
Turks and Caicos Is. (G.B.)	166	430	7,000
Virgin Is. (G.B.)	59	153	14,000
Virgin Is. (U.S.)	133	344	104,000
Wake I. (U.S.)	3	8	1,647
Wallis and Futuna Is. (Fr.)	106	275	9,000
Western Sahara§	102,703	266,000	165,000

†Populations are 1983 and earlier estimates based on the latest figures from official government and United Nations sources.
‡Claimed by Israel.
§Claimed by Morocco.

Africa

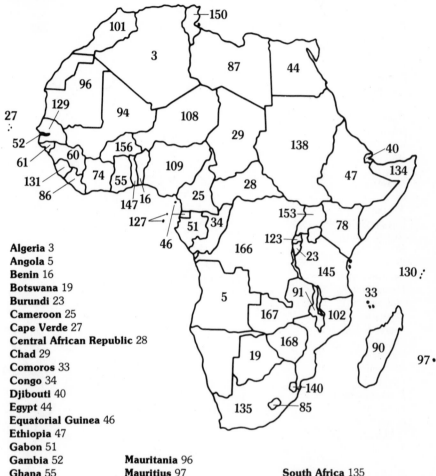

Algeria 3
Angola 5
Benin 16
Botswana 19
Burundi 23
Cameroon 25
Cape Verde 27
Central African Republic 28
Chad 29
Comoros 33
Congo 34
Djibouti 40
Egypt 44
Equatorial Guinea 46
Ethiopia 47
Gabon 51
Gambia 52
Ghana 55
Guinea 60
Guinea-Bissau 61
Ivory Coast 74
Kenya 78
Lesotho 85
Liberia 86
Libya 87
Madagascar 90
Malawi 91
Mali 94

Mauritania 96
Mauritius 97
Morocco 101
Mozambique 102
Niger 108
Nigeria 109
Rwanda 123
São Tomé and Príncipe 127
Senegal 129
Seychelles 130
Sierra Leone 131
Somalia 134

South Africa 135
Sudan 138
Swaziland 140
Tanzania 145
Togo 147
Tunisia 150
Uganda 153
Upper Volta 156
Zaire 166
Zambia 167
Zimbabwe (Rhodesia) 168

Asia

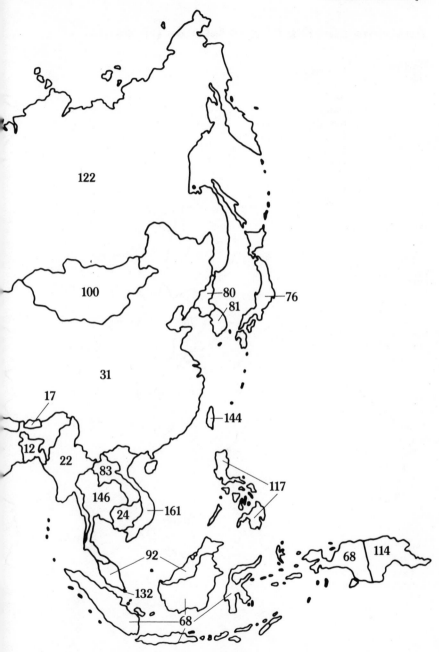

Australia and the Pacific Islands (Oceania)

Australia 8
Fiji 48
Kiribati 79
Nauru 103
New Zealand 106
Papua New Guinea 114
Solomon Islands 133
Tonga 148
Tuvalu 152
Vanuatu 158
Western Samoa 162

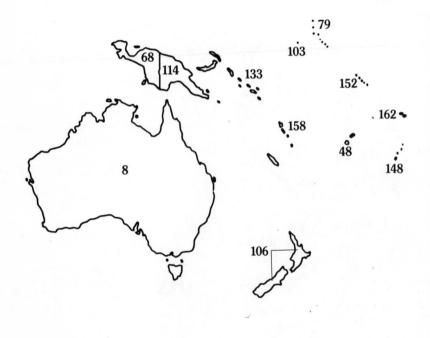

Europe

Albania 2
Andorra 4
Austria 9
Belgium 14
Bulgaria 21
Czechoslovakia 38
Denmark 39
Finland 49
France 50
Germany, East 53
Germany, West 54

Great Britain 56
Greece 57
Hungary 65
Iceland 66
Ireland 71
Italy 73
Liechtenstein 88
Luxembourg 89
Malta 95
Monaco 99
Netherlands 105
Norway 110
Poland 118

Portugal 119
Romania 121
Russia (European) 122
San Marino 126
Spain 136
Sweden 141
Switzerland 142
Vatican City 159
Yugoslavia 165

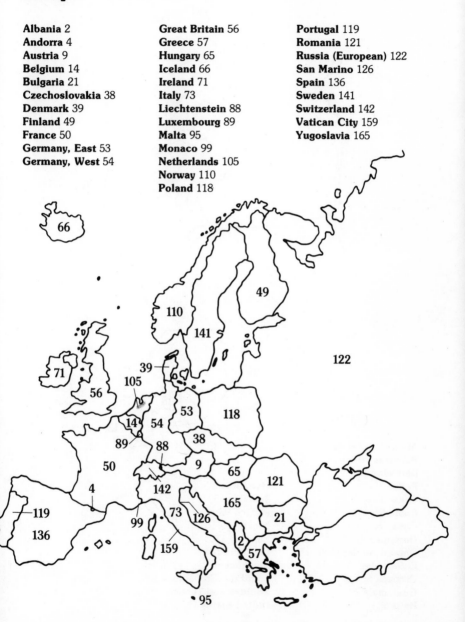

North America

South America

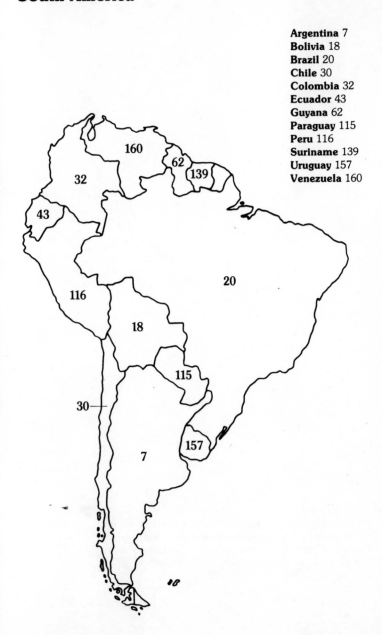

Argentina 7
Bolivia 18
Brazil 20
Chile 30
Colombia 32
Ecuador 43
Guyana 62
Paraguay 115
Peru 116
Suriname 139
Uruguay 157
Venezuela 160